Routing Techniques in Ad Hoc and Sensor Networks

Routing Techniques in Ad Hoc and Sensor Networks

Edited by **Frank Elliott**

WILLFORD **P**RESS

New York

Published by Willford Press,
118-35 Queens Blvd., Suite 400,
Forest Hills, NY 11375, USA
www.willfordpress.com

Routing Techniques in Ad Hoc and Sensor Networks
Edited by Frank Elliott

International Standard Book Number: 978-1-68285-111-1 (Hardback)

The publisher's policy is to use permanent paper from mills that operate a sustainable forestry policy. Furthermore, the publisher ensures that the text paper and cover boards used have met acceptable environmental accreditation standards.

Trademark Notice: Registered trademark of products or corporate names are used only for explanation and identification without intent to infringe.

Printed in the United States of America.

Contents

Preface

Routing may be applied to diverse set of networks such as telephones, internet, etc. Implementation of efficient routing techniques ensures hassle free transmission of data. This book provides a sufficient grounding in the field of routing techniques in ad hoc and sensor networks through detailed discussions on topics of significance, like distributed systems, collaborative networks, networking tools, semantic web, etc. This book, with its detailed analyses and data, will prove immensely beneficial to professionals and students involved in this area at various levels.

The researches compiled throughout the book are authentic and of high quality, combining several disciplines and from very diverse regions from around the world. Drawing on the contributions of many researchers from diverse countries, the book's objective is to provide the readers with the latest achievements in the area of research. This book will surely be a source of knowledge to all interested and researching the field.

In the end, I would like to express my deep sense of gratitude to all the authors for meeting the set deadlines in completing and submitting their research chapters. I would also like to thank the publisher for the support offered to us throughout the course of the book. Finally, I extend my sincere thanks to my family for being a constant source of inspiration and encouragement.

Editor

Performance of MC-MC CDMA Systems With Nonlinear Models of HPA

Labib Francis Gergis

Misr Academy for Engineering and Technology

Mansoura, Egypt

IACSIT Senior Member, IAENG Member

drlabeeb@yahoo.com

Abstract

A new wireless communication system denoted as Multi-Code Multi-Carrier CDMA (MC-MC CDMA), which is the combination of Multi-Code CDMA and Multi-Carrier CDMA, is analyzed in this paper. This system can satisfy multi-rate services using multi-code schemes and muti-carrier services used for high rate transmission. The system is evaluated using Traveling Wave Tube Amplifier (TWTA). This type of amplifiers continue to offer the best microwave high power amplifiers (HPA) performance in terms of power efficiency, size and cost, but lag behind Solid State Power Amplifiers (SSPA's) in linearity. This paper presents a technique for improving TWTA linearity. The use of pre-distorter (PD) linearization technique is described to provide TWTA performance comparable or superior to conventional SSPA's. The characteristics of the PD scheme is derived based on the extension of Saleh's model for HPA.

Keywords
Multi-Carrier CDMA, Multi-Code CDMA, MC-MC CDMA, TWTA, HPA, AM/AM, AM/PM, Saleh Model.

1. Introduction

Multi-Code CDMA and Multi-Carrier CDMA have attracted a lot of attention from researchers due to their perceived high rate transmission capability. In Multi-Code CDMA, researchers have investigated the systems performance in different fading channel [1] and suggested many schemes to improve the performance [2]. In Multi-Code CDMA, the input data streams are first split into several substreams in parallel and then orthogonal codes are multiplied for each substream.

In Multi-Carrier CDMA [3], the input data streams are first split into several substreams in parallel, like in Multi-Code CDMA and then modulate several subcarriers with each substream before transmitting the signals. Similarly with Multi-Code CDMA, Multi-Carrier CDMA is analyzed with different fading channels, and researchers have suggested schemes to improve the system performance [4], and [5].

In [6], [7], [8], and [9] Multi-Code Multi-Carrier CDMA system was evaluated and compared with both single code multi-carrier CDMA system and multi-code CDMA system with single carrier in a frequency selective fading channel.

Power amplifiers (PA's) are vital components in many communication systems. The linearity of a PA response constitutes an important factor that ensures signal integrity and reliable performance of the communication system. High power amplifiers (HPA) in microwave range suffer from the effects of amplitude modulation to amplitude modulation distortion (AM/AM), and amplitude modulation to phase modulation distortion (AM/PM), during conversions caused by the HPA amplifiers. These distortions can cause intermodulation (IM) distortion, which is undesirable to system designs. The effects of AM/AM and AM/PM distortions can cause the bit error rate performance of a communication channel to be increased.

The amplitude and phase modulation distortions are minimized using linearization methods. The linearization method requires modeling the characteristics of the amplitude distortion and phase distortion of the HPA. A Saleh model has been used to provide the linearization method and applied to measured data from HPA that characterize the distortion caused by the HPA. The measured data provides a performance curve indicating nonlinear distortion. The forward Saleh model is a math equation that describes the amplitude and phase modulation distortions of the HPA [10], and [11]. The amount of desired pre-distortion (PD) linearization is then determined to inversely match the amount of distortion for canceling out the distortion of the HPA.

The remainder of the paper is organized as follows. Section 2 discusses the proposed MC-MC CDMA system model. The bit error probability of the proposed system is derived in Section 3. In section 4, numerical results were presented, showing an improvement in the performance of MC-MC CDMA transmission system over the nonlinear channels. Section 5, summarized the conclusions had obtained through this paper.

2. System Model

2.1 Transmitter Model

The transmitter of the system model shown in Figure 1, is made up of two parts: the multi-code part and the multi-carrier part.

The multi-code part converts serial input data streams into parallel substreams, spreads each parallel substream to produce code division multiplexed bits with multi-codes, which are orthogonal to each other. Then, all substreams are summated to produce super-stream, $B_k(t)$.

In multi-carrier part, the super stream is serial-to-parallel (S/P) converted again, spread with a user specified Pseudo-random noise (PN) sequence, and modulated with orthogonal multi-carriers. Finally, the signal which is summated in the multi-carrier part is transmitted by the transmitter.

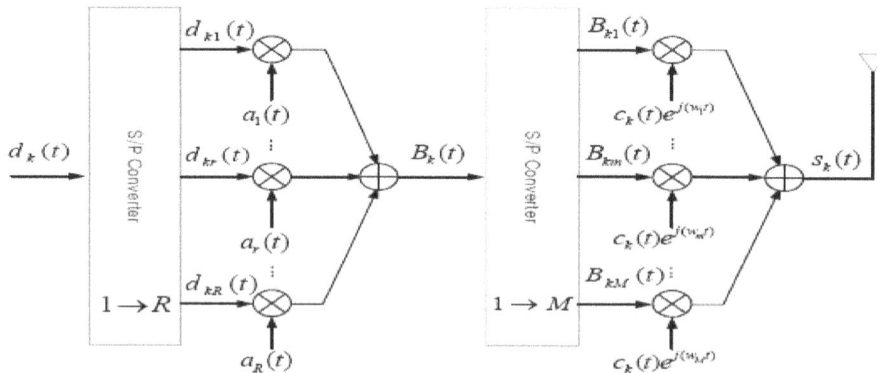

Figure 1. Transmitter Structure for the MC-MC CDMA System

By considering the system in which there are K users transmitting information simultaneously in cellular system. The transmitter for the kth using BPSK modulation, is shown in Figure 1. The data signal $d_k(t)$ is a random complex sequence representing the data bit and having a symbol rate of (RM / T), where R is the number of substreams in the multi-code, and T is the bit duration per input data stream, $d_k(t)$ can be expressed as

$$d_k(t) = d_k \, P_x(t) \tag{1}$$

where $x = T / RM$

After S/P conversion of the input data stream and modulating with orthogonal codes, the resulting output is given by

$$B_k(t) = \sum_{r=1}^{R} d_{kr}(t) \, a_r(t) \tag{2}$$

where $d_{kr}(t)$ is a substream S/P converted from the input data stream and is given by

$$d_{kr}(t) = d_{kr} \, P_{T/m}(t) \tag{3}$$

and a_r is an orthogonal code set for the rth substream of the R substreams, defined by

$$a_r(t) = a^n_r \, P_{TNa}(\, t - n \, T_{Na} \,) \tag{4}$$

where T_{Na} is the ship duration, N_a is the code length of the orthogonal codes, a^n_r is the nth value $\in \{\pm 1\}$ of the code a_r.

The transmitted signal $S_k(t)$ as a result of the summation of parallel super-substreams that S/P converted from the super-stream, multiplied by each subcarrier, and spread by the PN sequence. This is given by [6]

$$S_k(t) = \sqrt{2 \, P_k} \, \sum_{m=1}^{M} Re \{ \, B_{km}(t) \, C_k(t) \, e^{j(wmt)} \, \}$$

$$S_k(t) = \sqrt{2 \, P_k} \sum_{m=1}^{M} \sum_{r=1}^{R} Re \{ \, d_{krm}(t) \, a_r(t) \, C_k(t) \, e^{j(wmt)} \, \} \tag{5}$$

where

a) P_k is the signal power of user k distributed among the carriers, assuming $P_1 = P_2 = \ldots = P_k = P$.

b) $B_{km}(t)$ is the mth super-substream S/P converted from super-stream $B_k(t)$ with a bit rate of (R/T). After S/P conversion, the symbol duration increases M times.

c) $d_{krm}(t)$ is the data symbol of rth substream of the mth super-stream with value

$$d_{krm}(t) = d_{krm} \, P_T(t) \, .$$

d) $C_k(t)$ is the PN sequence defined as

$$N_c - 1$$

$$C_k(t) = \sum_{s=0} C^s_k \; P_{TNc} \, (t - s \, T_{Nc}) \tag{6}$$

$$T_{Nc} = T / N_c$$

N_c is the length of a PN sequence and C^s_k (t) is the sth value of the PN sequence. The subcarrier $e^{j(w_m)}$ is the mth subcarrier having frequency f_m, defined by

$$w_m = 2\pi f_m \, , f_m = R_m / T \, , \text{ and } m = 1, 2, \ldots\ldots, M. \tag{7}$$

R is the number of substream.

2.2 Nonlinear Model

High power amplifiers exist in almost all wireless communication links. Due to various nonlinear electronic components inside them, these power amplifiers are nonlinear devices. Power amplifiers exhibit nonlinear distortion in both amplitude and phase. The amplitude conversion is referred as Amplitude Modulation to Amplitude Modulation (AM/AM) conversion, and phase conversion is reffered as Amplitude Modulation to Phase Modulation (AM/PM) conversion.

A major type of power amplifiers typically used in communication systems, is travelling wave tube amplifiers (TWTA).

The classical and most often used nonlinear model of power amplifier is Saleh's model [10]. It is a pure nonlinear model without memory. The output of HPA defined in Figure .2, is expressed as

$$S_y = A \, [\, U_x \,] \, e^{j(\alpha x + \, \Phi[Ux])} \tag{8}$$

where the input-output functional relation of the HPA has been defined as a *transfer function*.

The equations define this base-band model of HPA as two modulus dependent *transfer functions* are defined as [9] and [10] :

$$A[U_x] = \alpha_a \, U_x \; / \; 1 + \beta_a \, U^2_x$$

$$\Phi[U_x] = \alpha_\Phi \, U_x \; / \; 1 + \beta_\Phi \, U^2_x \tag{9}$$

where $A[U_x]$ and $\Phi[U_x]$ are the corresponding AM/AM and AM/PM characteristics respectively, both dependent exclusively on U_x , which is the input modulus to HPA.

The values of α_a, β_a , α_Φ and β_Φ are defined in [10]. The corresponding AM/AM and AM/PM curves so scaled are depicted in Figure 2.

The HPA operation in the region of its nonlinear characteristic causes a nonlinear distortion of a transmitted signal, that subsequently results in increasing the bit error rate (*BER*), and the out-of-band energy radiation (spectral spreading).

The operating point of HPA is defined by input back-off (*IBO*) parameter which corresponds to the ratio of saturated input power (P_{max}), and the average input power (P_{in}) [11] :

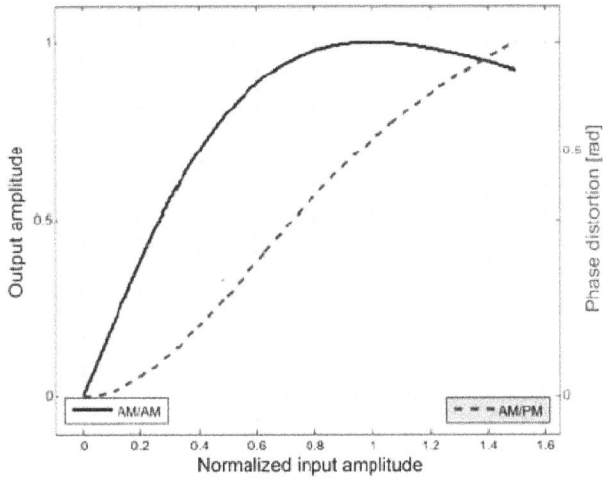

Figure 2. AM/AM and AM/PM characteristics of the normalized Saleh model

$$IBO_{dB} = 10 \log_{10} \quad (P_{max} / P_{in}) \qquad (10)$$

The measure of effects due to the nonlinear HPA could be decreased by the selection of relatively high values of *IBO*

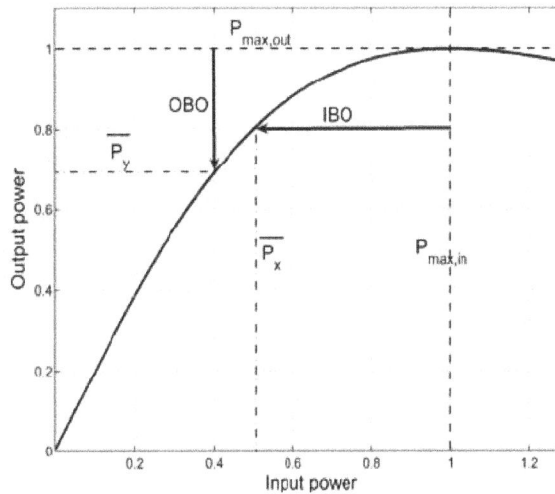

Figure. 3 Graphical representation of IBO and OBO.

2.3   Receiver Model

The receiver structure is shown in Figure 4, where signals consisting of *M* carriers are demodulated by a locally generated carriers, despread by a user specific PN sequence, and parallel-to-serial (P/S) converted to produce $\hat{B}_k(t)$. Then $B_k(t)$ is despread with a_r code sequence, correlated over one symbol period then P/S converted to recover $\hat{d}_{kr}$.

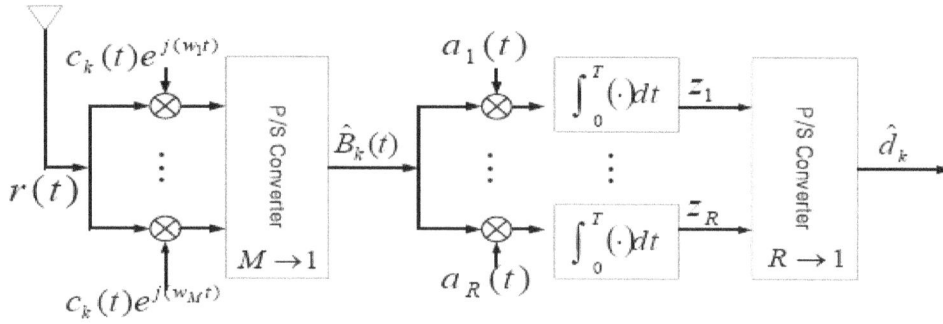

Figure 4. Receiver Structure for the MC-MC CDMA System

The overall received signal equation corrupted with Additive White Gaussian Noise (AWGN) is given by

$$r(t) = \sum_{k=1}^{k} y_k(t) + n(t)$$

$$= \sqrt{2P}\ A_{11}\ Re\ [\ d_{111}(t\text{-}\tau_{11})\ a_1(t\text{-}\tau_{11})c_1(t\text{-}\tau_{11})\ e^{j\{w_1(t\text{-}\tau_{11})+\Phi_{11}\}}\ ]$$

$$+ \sqrt{2P}\ \sum_{l=2}^{L} A_{11}\ Re\ [\ d_{111}(t\text{-}\tau_{ll})\ a_1(t\text{-}\tau_{ll})c_1(t\text{-}\tau_{ll})\ e^{j\{w_1(t\text{-}\tau_{ll})+\Phi_{ll}\}}\ ]$$

$$+ \sqrt{2P}\ \sum_{r=2}^{R}\sum_{l=1}^{L} A_{11}\ Re\ [\ d_{1r1}(t\text{-}\tau_{ll})\ a_r(t\text{-}\tau_{ll})c_1(t\text{-}\tau_{ll})\ e^{j\{w_1(t\text{-}\tau_{ll})+\Phi_{ll}\}}\ ]$$

$$+ \sqrt{2P}\ \sum_{m=2}^{M}\sum_{r=1}^{R}\sum_{l=1}^{L} A_{11}\ Re\ [\ d_{1rm}(t\text{-}\tau_{ll})\ a_r(t\text{-}\tau_{ll})c_1(t\text{-}\tau_{ll})\ e^{j\{w_m(t\text{-}\tau_{ll})+\Phi_{ll}\}}\ ]$$

$$+ \sqrt{2P}\ \sum_{K=2}^{K}\sum_{m=1}^{M}\sum_{r=1}^{R}\sum_{l=1}^{L} A_{kl}\ Re\ [\ d_{krm}(t\text{-}\tau_{ll})\ a_r(t\text{-}\tau_{kl})c_k(t\text{-}\tau_{kl})$$

$$\cdot\ e^{j\{w_m(t\text{-}\tau_{kl})+\Phi_{kl}\}}\ ]$$

$$+\ n(t) \tag{11}$$

The received signal can be expressed in six components as

$$r(t) = r_{DS}(t) + r_{MPI}(t) + r_{ISSI}(t) + r_{ICI}(t) + r_{MUI}(t) + n(t) \tag{12}$$

where
  a) $r_{DS}(t)$ is the desired signal... corresponding to reference user's $(k=1)$, reference substream $(r=1)$, reference carrier $(m=1)$, and reference path $(l=1)$.

b) $r_{MPI}(t)$ is the interference caused by the propagation of the desired signal and corresponds to reference user's ($k=1$), reference substream ($r=1$), and reference carrier ($m=1$), from other paths ($l \neq 1$), which is called Multipath Interference (MPI)

c) $r_{ISSI}(t)$ is the interference caused by other substreams except the reference substream, which is called Inter-Substream Interference (ISSI)

$r_{ICI}(t)$ is the interference caused by other carriers except the reference carrier, which is called Inter-Carrier Interference (ICI)

d) $r_{MUI}(t)$ is the interference caused by other users except the reference user ($k=1$), which is commonly known as Multi User Interference (MUI)

e) $n(t)$ is the AWGN component.

At the receiver, the received signal is first demodulated by locally generated carrier and then despread by a user specific code sequence before P/S conversion.

The output signals of the P/S is despread again by each orthogonal code for multicode component   in order to recover a substream before correlating over a period T. Finally R substreams are recovered
from the correlated outputs.

The output of the correlator also may be decomposed into six components

$$z_1 = z_{DS} + z_{MPI} + z_{ISSI} + z_{ICI} + z_{MUI} + z_n \tag{13}$$

$$z_1 = z_{DS} + I_{Total} \tag{14}$$

where each component is similarly defined as in (12), and z_n is the correlated AWGN component, and I_{Total} is the sum of the all interference terms and AWGN component.
The desired signal power can be defined from the first term of (13). Using $(d^l_{111})^2 = 1$

$$(z_{DS})^2 = S = (P/2)(A_{11})^2 T^2 \tag{15}$$

3. Bit Error Rate Performance (BER)

The average bit error probability can be expressed as [8] :

$$\overline{P_e} = \int_0^\infty f(A_{11}) \, P_e(A_{11}) \, dA \tag{16}$$

where $f(A_{11})$ is the pdf of random variable A_{11}.

$$P_e(A_{11}) = 1/2 \, erfc(\gamma) \tag{17}$$

where $erfc(.)$ is the complementary error function, and γ is defined as [9] :

$$\gamma = S / \sigma^2_{Total} = (P/2)(A_{11})^2 T^2 / \sigma^2_{Total} \tag{18}$$

σ^2_{Total} is the total variance for the all interference terms and AWGN component, it can be written as

$$\sigma^2_{Total} = \sigma^2_{MPI} + \sigma^2_{ISSI} + \sigma^2_{ICI} + \sigma^2_{MUI} + \sigma^2_n \tag{19}$$

4. Results and Discussions

In this section, the BER performance of MC-MC CDMA is derived and compared with Multi-Code CDMA system, and Multi-Carrier CDMA system, and some properties of MC-MC CDMA are observed. Fig. 5, shows the comparison assuming that all systems use BPSK modulation technique, number of substreams for Multi-Code CDMA and MC-MC CDMA (R) equals 8, number of carriers for Multi-Carrier CDMA and MC-MC CDMA (M) equals 8, and number of users (K) equals 20. The Figure indicates that the MC-MC CDMA system has the lowest BER performance of all the system compared.

Fig. 6, illustrates the BER performance of MC-MC CDMA *(R=M=8)* as a function of number of users *(K = 1, 10, and 50)*. It is shown that with increasing simultaneous users, BER increase, and the performance decreases.

Fig. 7 shows the effect of the number of subcarriers M on the BER performance of MC-MC CDMA system *(R=8, K=20)* . It is clear that by increasing M values, a higher BER performance is gained.

The BER performance of MC-MC CDMA system *(R=M=8, K=20)* is expressed under the case of nonlinearity distortions (AM/AM and AM/PM), for different values of *IBO (IBO = 7 dB* and *IBO = 9 dB)*, compared with the case of no nonlinear (that had linearized by PD), is illustrated in Fig. 8. It is shown that it is highly recommended to use linearization devices at the transmitter side in order to suppress the undesirable nonlinearity effects and to get improved bit error performance.

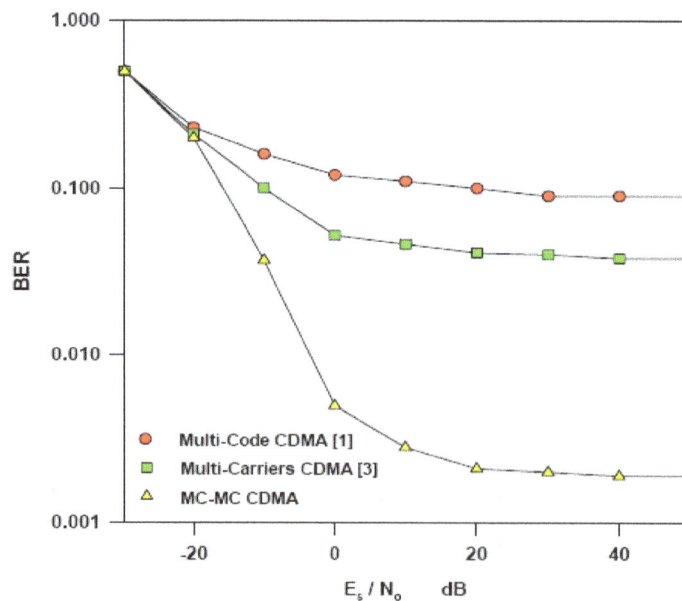

Figure 5. Bit Error Rate Respect to Multi-Code CDMA, Multi-Code CDMA

and MC-MC CDMA System

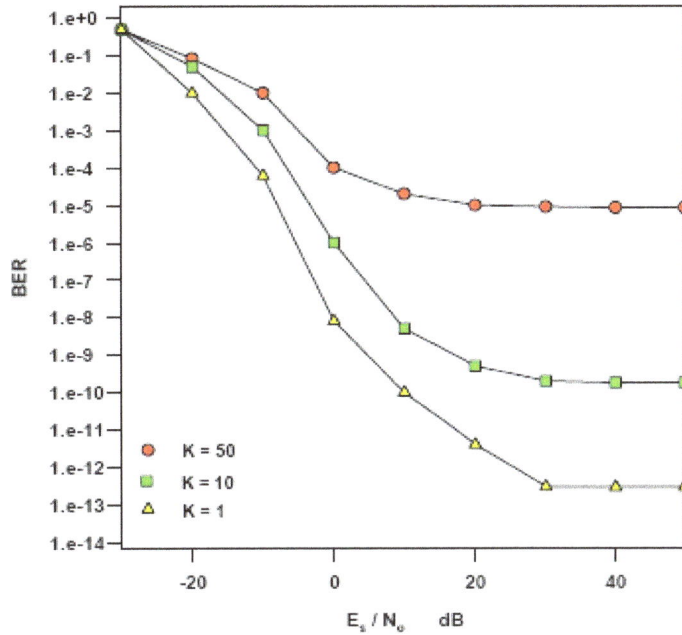

**Figure 6. BER Performance of MC-MC CDMA System as a Function
of Number of Users (K)**

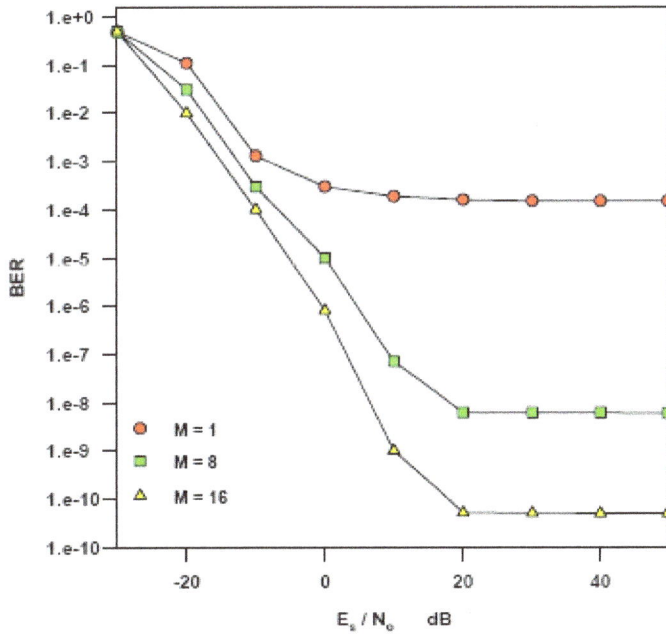

**Figure 7. BER Performance of MC-MC CDMA System as a Function
of Number of Subcarriers (M)**

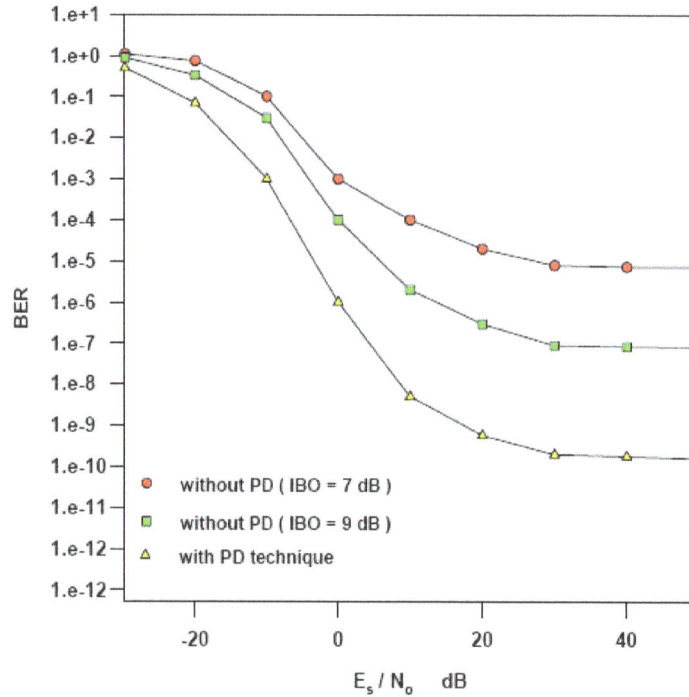

**Figure 8. Effect of Using Pre-Distorter Technique (PD)
on the BER Performance of MC-MC CDMA System**

5. CONCLUSIONS

In this paper, the analysis and performance results of MC-MC CDMA system is presented in terms of BER. The performance of the MC-MC CDMA is compared to the performance of Multi-Carrier CDMA, and Multi-Code CDMA. It was shown that the MC-MC CDMA system outperformed other systems.

The effects of nonlinearities (AM/AM and AM/PM) were analyzed. It is shown that these effects can be compensated by using PD technique. From analytical results, it is confirmed that PD system with MC-MC CDMA gives a good performance improvements compared to MC-MC CDMA signals without PD.

References

[1] J. Iong, and Z. Chen, " Theoretical Analysis of an MC-CDMA System in Cellular Environments", Journal of Science and Engineering Technology, *VOL. 6, No. 1, pp. 37-50.* 2010

[2] M. Frikel, S. Safi, B. Targui, and M. Saad, " Channel Identification Using Chaos for an Uplink/Downlink Multicarrier Code Division Multiple Access System ", Journal of Telecommunications and Information Technology, pp. 48-54, 1/ 2010.

[3] P. H. Lien and L. N. Anh," PMC-MC-CDMA and MMC-MC-CDMA," International Symposium on Electrical & Electronics Engineering, HCM City, Vietnam, October 2007.

[4] J. Iong and Z. Chen, " Performance Analysis of MC-CDMA Communication Systems over Nakagami-M Environments," Journal of Marine Science and

Technology, Vol. 14, No. 1, pp. 58-63, 2006.

[5] P. Harinath, and V. Reddy, " BER Degradation of MC-CDMA at High SNR with MMSE Equalization and Reidual Frequency Offset ", EURASIP Journal on Wireless Communication and Networking, Article ID 293264, Volume 2009.

[6] S. Praveen, N. Nagarajan, and V. Arthi, " Selection Based Succesive Interference Cancellation for Multicode Multicarrier CDMA Transceiver ", WSEAS Transactions on Communications , Issue 8, Volume 9, August 2010.

[7] E. Kunnari, " MULTIRATE MC-CDMA .. Performance analysis in stochastically modeled correlated fading channels, with an application to OFDM-UWB ", OULU UNIVERSITY PRESS, 2008.

[8] T. Kim, J. G. Andrews, J. Kim, and T. Rappaport," Multi-Code Multicarrier CDMA : Performance Analysis," IEEE International Conference on Communications (ICC' 2004), Paris, France, June 2004.

[9] M. A. Zahieh, " Design and Analysis of Multicarrier Multicode Wavelet Packets Based CDMA Communications Systems with Multiuser Detection," Ph.D Thesis.. University of Akron, August 2006.

[10] S. Chang, " An efficient compensation of TWTA's nonlinear distortion in wideband OFDM system," IEICE Electronics Express, Vol. 6, No. 2, pp. 111-116, 2009.

[11] I. Teikari, " Digital Predistortion Linearization Mothods for RF Power Amplifiers ", Doctoral Dissertation, Helsinki University of Technology, 2008.

PERFORMANCE EVALUATION OF AODV AND DSR ON-DEMAND ROUTING PROTOCOLS WITH VARYING MANET SIZE

Nilesh P. Bobade[1], Nitiket N. Mhala[2]

[1]Department of Electronics Engineering, Bapurao Deshmukh COE, Sevagram, Wardha, M.S., India.
b_nilesh246@rediffmail.com
[2]Department of Electronics Engineering, Bapurao Deshmukh COE, Sevagram, Wardha, M.S., India.
nitiket_m@rediffmail.com

ABSTRACT

A mobile ad hoc network (MANET) is a collection of wireless mobile nodes dynamically forming a network topology without the use of any existing network infrastructure or centralized administration. Routing is the process which transmitting the data packets from a source node to a given destination. The main procedure for evaluating the performance of MANETs is simulation. The on-demand protocol performs better than the table-driven protocol. Different methods and simulation environments give different results. It is not clear how these different protocols perform under different environments. One protocol may be the best in one network configuration but the worst in another. In this paper an attempt has been made to compare the performance of on demand reactive routing protocols i.e. Ad hoc On Demand Distance Vector (AODV) and Dynamic Source Routing (DSR). As per our findings the differences in the protocol mechanics lead to significant performance differentials for both of these protocols. Always the network protocols were simulated as a function of mobility, but not as a function of network density. In our paper the performance of AODV and DSR is evaluated with respect to performance metrics like Packet Delivery Fraction (PDF), Average end-to-end delay, Normalized Routing Load (NRL) and throughput by varying network size up to 50 nodes. These simulations are carried out using the NS-2 which is the main network simulator, NAM (Network Animator), AWK (post processing script). Our results presented in this research work demonstrate the concept AODV and DSR routing protocols w.r.t. MANET size in an Ad hoc environment.

KEYWORDS

MANET, AODV, DSR, Performance Metrics, NS-2.34& Simulation

1. INTRODUCTION

A Mobile Ad hoc Network (MANET) is a system of wireless mobile nodes which can freely and dynamically self-organize and co-operative in to arbitrary and temporary network topologies, allowing peoples and devices to communicate without any pre-existing communication architecture. Each node in the ad hoc network acts as a router, forwarding data packets for other nodes. A central challenge in the design of mobile ad hoc networks is the development of routing protocols that can efficiently find the transmission paths between two communicating nodes. The ad hoc networks are very flexible and suitable for several types of applications due to its feature like they allow the establishment of temporary communication without any pre-installed infrastructure. With newly emerging radio technologies, e.g. IEEE 802.11and Bluetooth, the realization of multimedia applications over mobile ad-hoc networks becomes more realistic. Our goal is to carry out a systematic performance study of an on demand routing protocol AODV [1,

14] and DSR [1] for ad hoc networks. However our performance evaluation is based on varying node density in the Mobile ad hoc Network. Generally the network protocols were simulated as a function of pause time (node mobility), but not as a function of network size. The rest of the paper is organized as follows: The related work is provided in section 2. The AODV and DSR routing protocol Description are summarized in section 3 and 4 resp. The simulation environment and performance metrics are described in Section 5. We present the simulation results and observation in section 6 and the conclusion is presented in section 7.

2. RELATED WORK

Several researchers have done the quantitative and qualitative analysis of Ad hoc Routing Protocols by means of different performance parameters. Also they have used different simulators for this purpose.

1) J Broch et al. [1] performed experimental performance comparison of both proactive and reactive routing protocols. In their NS-2 simulation, a network density of 50 nodes with varying pause times and various movement patterns were chosen.

2) Jorg D.O. [3] studied the behavior of different routing protocols for the changes of network topology which resulting from link breaks, node movement, etc. In his paper, performance of routing protocols was evaluated by varying number of nodes. But he did not investigate the performance of protocols under high mobility, large number of traffic sources and larger number of nodes in the network which may lead to congestion situations.

3) Khan et al. [4] studied and compared the performance of routing protocols by using NCTUns network simulator. In their paper, performance of routing protocols was evaluated by varying number of nodes in multiples of 5 in the ad hoc network. The simulations were carried out for 70 seconds of the simulation time. The packet size was fixed to 1400 bytes.

4) Arunkumar B R et al. Authors perform simulations by using NS-2 simulator [13]. Their studies have shown that reactive protocols perform better than table driven (proactive) protocols.

5) S. Gowrishanker et al [9] performed the analysis of OLSR and AODV by using NS-2, the simulation period for each scenario was 900 seconds and the simulated mobility network area was 800 m x 500 m. In each simulation scenario, the nodes were initially located at the center of the simulation region. The nodes start moving after the first 10 seconds of simulated time. In it, the application used to generate is CBR traffic and IP is used as Network layer protocol.

6) N Vetrivelan & Dr. A V Reddy [10] analyzed the performance differentials using varying network density and simulation times. They performed two simulation experiments for 10 & 25 nodes with simulation time up to 100 sec.

7) S. P. Setty et.al. [6] evaluated the performance of existing wireless routing protocol AODV in various nodes placement models like Grid, Random and Uniform using QualNet 5.0.

3. AODV ROUTING PROTOCOL DESCRIPTION

Ad hoc On Demand Distance Vector (AODV) [14] is a reactive routing protocol which initiates a route discovery process only when it has data packets to transmit and it does not have any route path towards the destination node, that is, route discovery in AODV is called as on-demand. AODV uses sequence numbers maintained at each destination to determine freshness of routing information and to avoid the routing loops that may occur during the routing calculation process. All routing packets carry these sequence numbers.

3.1. Route Discovery Process

During a route discovery process, the source node broadcasts a route query packet to its neighbors. If any of the neighbors has a route to the destination, it replies to the query with a route reply packet; otherwise, the neighbors rebroadcast the route query packet. Finally, some query packets reach to the destination.

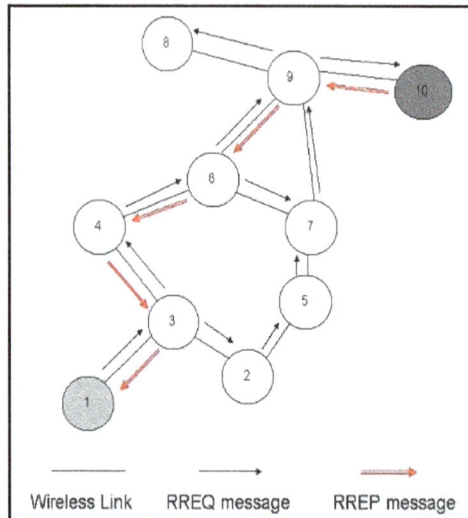

Figure 1. AODV Route Discovery Process

Figure 1 shows the route discovery process from source node1 to destination node 10. At that time, a reply packet is produced and transmitted tracing back the route traversed by the query packet as shown in Figure 1.

3.2. AODV Route Message Generation

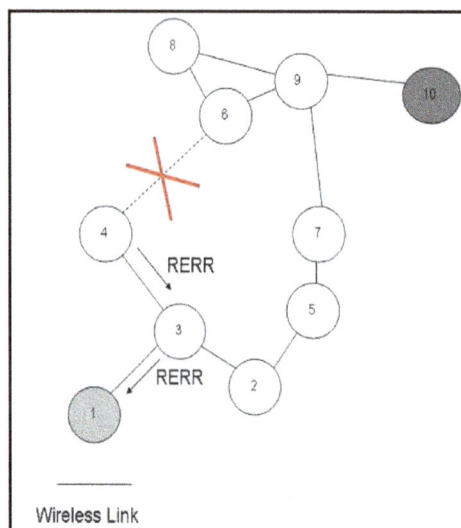

Figure 2. AODV Route Error message generation

The route maintenance process in AODV is very simple. When the link in the communication path between node 1 and node 10 breaks the upstream node that is affected by the break, in this case node 4 generates and broadcasts a RERR message. The RERR message eventually ends up in source node 1. After receiving the RERR message, node 1 will generate a new RREQ message (Figure 2).

3.3. AODV Route Maintenance Process

Finally, if node 2 already has a route to node 10, it will generate a RREP message, as indicated in Figure 3. Otherwise, it will re-broadcast the RREQ from source node 1 to destination node 10 as shown in Figure 3.

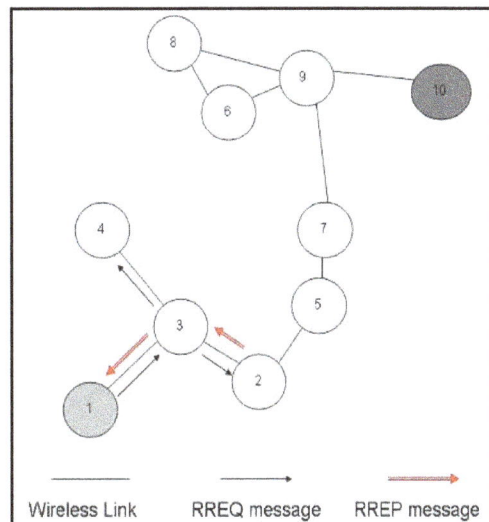

Figure 3. AODV Route Maintenance Process

4. DSR Routing Protocol Description

The Dynamic Source Routing (DSR) protocol is a reactive routing protocol based on source routing. In the source routing, a source determines the perfect sequence of nodes with which it propagate a packet towards the destination. The list of intermediate nodes for routing is explicitly stored in the packet's header.

In DSR, every mobile node needs to maintain a route cache where it caches source routes. When a source node wants to send a packet to some other intermediate node, it first checks its route cache for a source route to the destination for successful delivery of data packets. In this case if a route is found, the source node uses this route to propagate the data packet otherwise it initiates the route discovery process. Route discovery and route maintenance are the two main features of the DSR protocol.

4.1. Route Discovery

For route discovery, the source node starts by broadcasting a route request packet that can be received by all neighbor nodes within its wireless transmission range. The route request contains the address of the destination host, referred to as the target of the route discovery, the source's address, a route record field and a unique identification number (Figure 4). At the end, the source node should receive a route reply packet with a list of network nodes through which it should transmit the data packets that is supposed the route discovery process was successful [3,16].

During the route discovery process, the route record field is used to contain the sequence of hops which already taken. At start, all senders initiate the route record as a list with a single node containing itself. The next intermediate node attaches itself to the list and so on. Each route request packet also contains a unique identification number called as request_id which is a simple counter increased whenever a new route request packet is being sent by the source node. So each route request packet can be uniquely identified through its initiator's address and request_id. When a node receives a route request packet, it is important to process the request in the following given order. This way we can make sure that no loops will occur during the broadcasting of the packets.

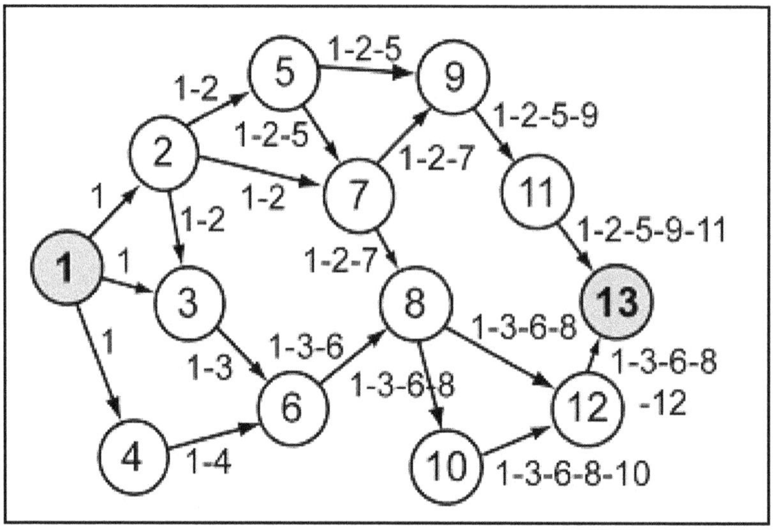

Figure 4. Building of the record during route discovery in DSR

□ If the pair < source node address, request_id > is found in the list of recent route requests, the packet is discarded.

□ If the host's address is already listed in the request's route record, the packet is also discarded. This indicates removal same request that arrive by using a loop.

□ If the destination address in the route request matches the host's address, the route record field contains the route by which the request reached this host from the source node. A route reply packet is sent back to the source node with a copy of this route.

□ Otherwise, add this node's address to the route record field and re-broadcast this packet.

A route reply is sent back either if the request packet reaches the destination node itself, or if the request reaches an intermediate node which has an active route4 to the destination in its route cache. The route record field in the request packet indicates the sequence of hops which was considered. If the destination node generating the route reply, it just takes the route record field of the route request and puts it into the route reply. If the responding node is an intermediate node, it attaches the cached route to the route record and then generates the route reply (Figure 5).

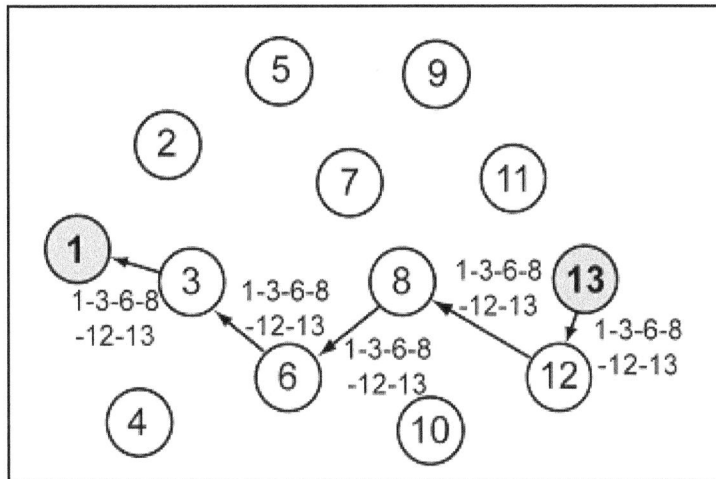

Figure 5. Propagation of the route reply in DSR

Sending back route replies can be processed with two different ways: DSR may use symmetric links. In the case of symmetric links, the node generating the route reply just uses the reverse route of the route record. When using asymmetric links, the node needs to initiate its own route discovery process and back the route reply on the new route request.

4.2. Route Maintenance

Route maintenance can be accomplished by two different processes:

☐   Hop-by-hop acknowledgement at the data link layer
☐   End-to-end acknowledgements

Hop-by-hop acknowledgement is the process at the data link layer which allows an early detection and re-transmission of lost packets. If the data link layer determines a fatal transmission error, a route error packet is being sent back to the sender of the packet. The route error packet contains the information about the address of the node detecting the error and the host's address which was trying to transmit the packet. Whenever a node receives a route error packet, the hop is removed from the route cache and all routes containing this hop are truncated at that point.

When wireless transmission between two hosts does not process equally well in both directions, end-to-end acknowledgement may be used. As long as a route exists, the two end nodes are able to communicate and route maintenance is possible. In this case, acknowledgements or replies on the transport layer used to indicate the status of the route from one host to the another. However, with end-to-end acknowledgement it is not possible to find out the hop which has been in error.

5. SIMULATION ENVIRONMENT

5.1. Simulation Model

Here we give the significance for the evaluation of performance of Ad Hoc routing protocol AODV with varying the number of mobile nodes. The network simulations have been done using network simulator NS-2 [13]. The network simulator NS-2 is discrete event simulation software for network simulations which means it simulates events such as sending, receiving, forwarding and dropping packets. The latest version, ns-allinone-2.34, supports simulation for routing protocols for ad hoc wireless networks such as AODV, DSDV, TORA, and DSR. NS-2 is written in C++ programming language with Object Tool Common Language (OTCL). Although NS-2. 34 can be built on different platforms, for this paper, we chose a Linux platform i.e. FEDORA 7, as Linux offers a number of programming development tools that can be used with the simulation process. To run a simulation with NS-2.34, the user must write the OTCL simulation script. We get the simulation results in an output trace file and here, we analyzed the experimental results by using the awk command (Figure 8 & 9).The performance parameters are graphically visualized in XGRAPH v12.1(Figure 10, 11, 12 & 13). NS-2 also offers a visual representation of the simulated network by tracing nodes movements and events and writing them in a network animator (NAM) file (Figure 6 & 7).

5.2. Simulation Parameters

In our work, we consider a network of nodes placing within a 1000m X 1000m area. The performance of AODV and DSR is evaluated by keeping the network speed and pause time constant and varying the network size (number of mobile nodes). Table 1 shows the simulation parameters used in this evaluation.

Table 1. Parameters values for AODV and DSR Simulation

Simulation Parameters	
Simulator	NS-2.34
Protocols	**AODV and DSR**
Simulation duration	200 seconds
Simulation area	1000 m x 1000 m
Number of nodes	**5, 10, 15, 20, 25, 30, 35, 40, 45, 50**
Transmission range	250 m
Movement model	Random Waypoint
MAC Layer Protocol	IEEE 802.11
Pause Time	100 sec
Maximum speed	20 m/s
Packet rate	4 packets/sec
Traffic type	CBR (UDP)
Data Payload	512 bytes/packet

5.3. Performance Metrics

While analyzed the AODV and DSR protocols, we focused on four performance metrics for evaluation which are Packet Delivery Fraction (PDF), Average End-to-End Delay, Normalized Routing Load (NRL) and Throughput.

5.3.1. Packet delivery fraction

Packet delivery fraction (PDF) is the fraction of all the received data packets successfully at the destinations over the number of data packets sent by the CBR sources.

5.3.2. Average End to end delay

It is the average time from the transmission of a data packet at a source node until packet delivery to a destination which includes all possible delays caused by buffering during route discovery process, retransmission delays, queuing at the interface queue, propagation and transfer times of data packets.

5.3.3. Normalized Routing Load

The normalized routing load (NRL) is as the ratio of all routing control packets sent by all nodes to the number of received data packets at the destination nodes.

5.3.4. Throughput

It is the average number of messages successfully delivered per unit time or it is the average number of bits delivered per second.

6. SIMULATION RESULTS & OBESRVATION

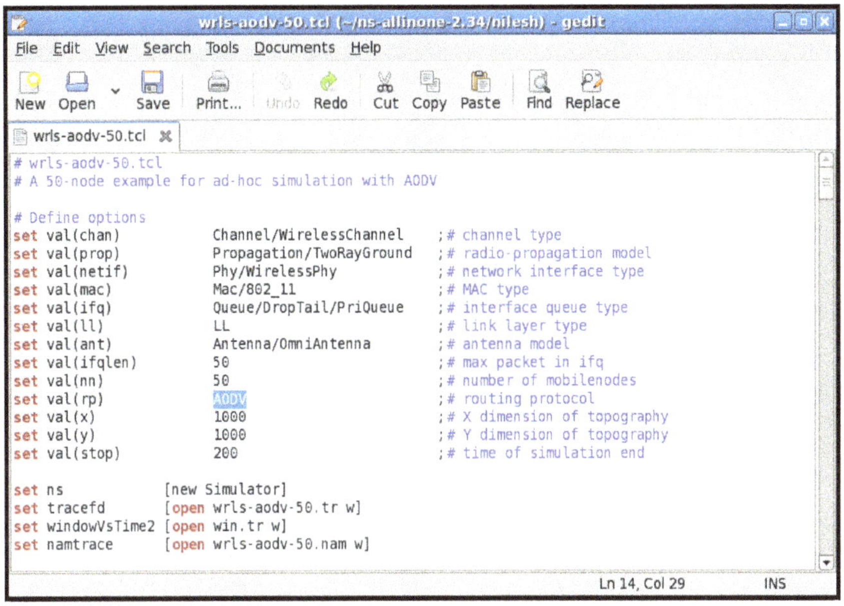

Figure 6.   Screenshot of AODV Tcl script

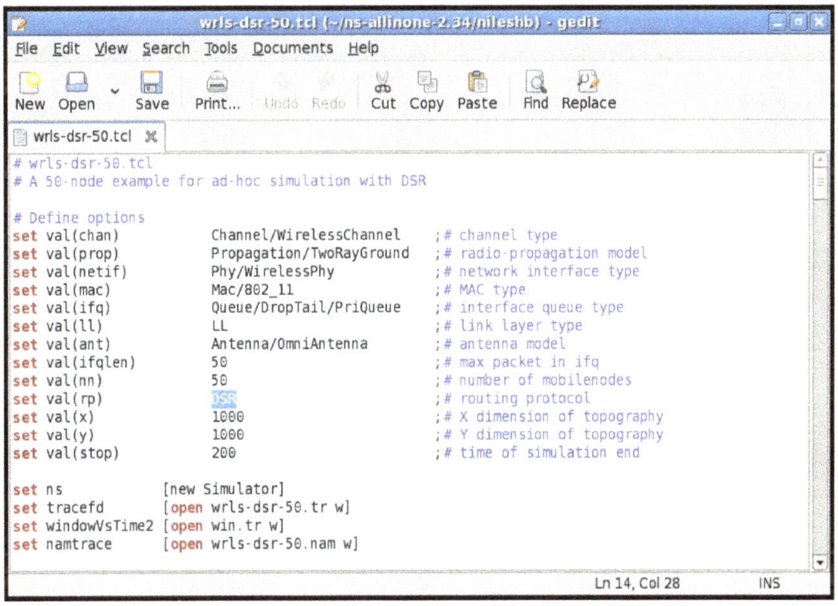

Figure 7. Screenshot of DSR Tcl script

Figure 6 and figure 7 show the screenshots of AODV and DSR Tcl script. The results after simulation are viewed in the form of line graphs. The performance of AODV and DSR based on the varying the network size i.e. no. of nodes is done on parameters like packet delivery fraction, average end-to-end delay, normalized routing load and throughput.

Figure 8 and figure 9 show the creation of clusters with 50 mobile nodes for AODV and DSR respectively as it is shown in the NAM console which is a built-in program in NS-2-allinone package after the end of the simulation process.

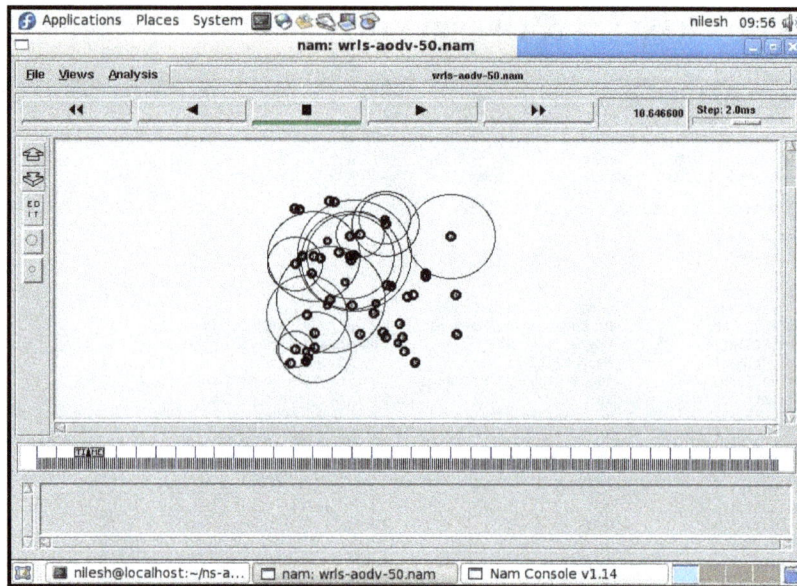

Figure 8. AODV with 50 nodes: Route Discovery

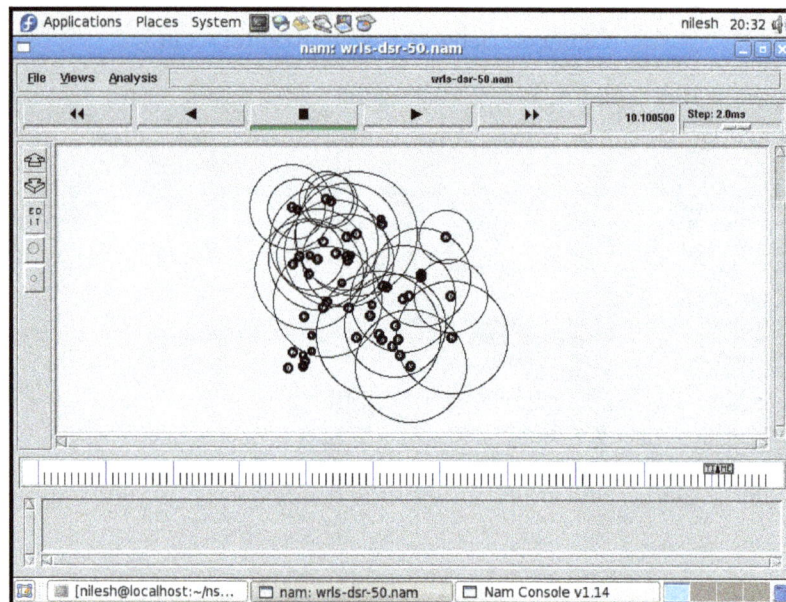

Figure 9. DSR with 50 nodes: Route Discovery

Figure 10 and 11 shows the calculation of send packets, received packets, packet delivery fraction, average end-to-end delay, normalized routing load and etc. for AODV and DSR simulation resp. (50 nodes) by running AWK script for it.

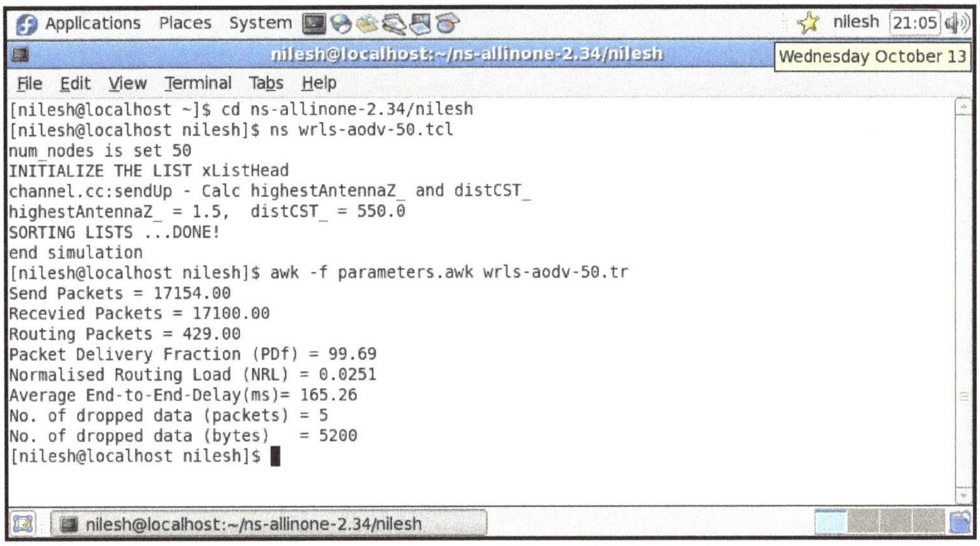

Figure 10. Screenshot of the results of performance metrics for AODV simulation

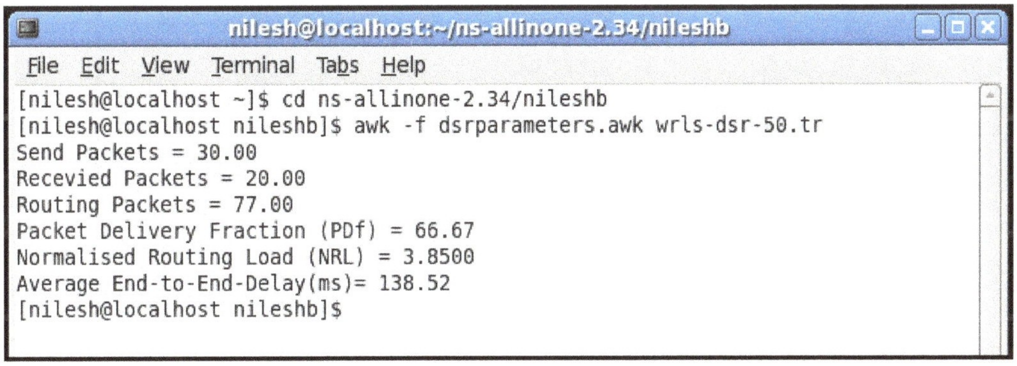

Figure 11. Screenshot of the results of performance metrics for DSR simulation

Figure 12 highlights the relative performance of AODV and DSR. When looking at the packet delivery ratio, it can easily be seen that AODV perform much better than DSR. AODV delivers a greater percentage of the originated data i.e. almost 100%. The low packet delivery fraction of DSR may be explained by the aggressive route caching built into this protocol. Further it is observed that the performance of AODV is consistently uniform between 99.5 % & 99.7 %.

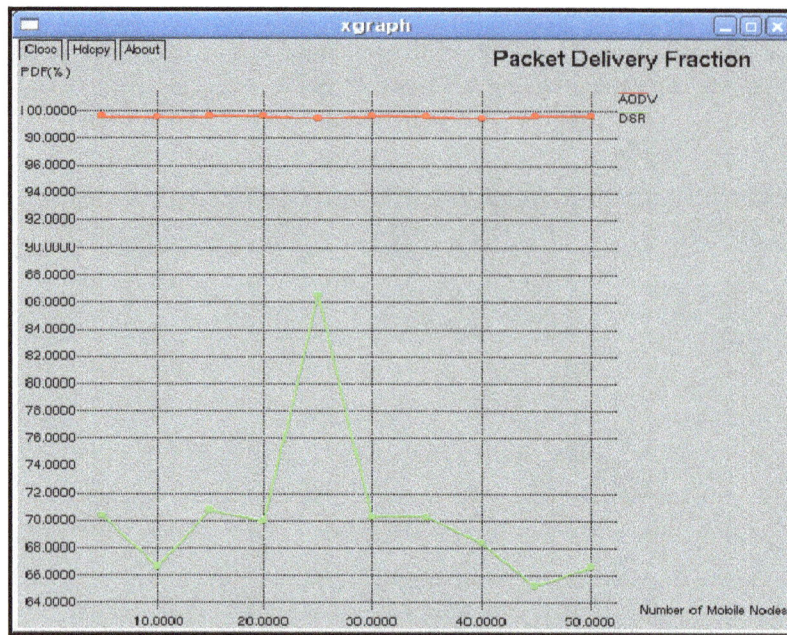

Figure 12. Packet Delivery Fraction for AODV and DSR with
varying no. of Mobile Nodes

From figure 13, it is clear that the average delay of AODV is higher than DSR. The performance of AODV is almost uniform (below 180 ms) except for 40 nodes.

Figure 13. Average End-to-End Delay for AODV and DSR with
varying no. of Mobile Nodes

From figure 14, we can observe that AODV demonstrates significantly lower routing load than DSR. It is almost the consistent.

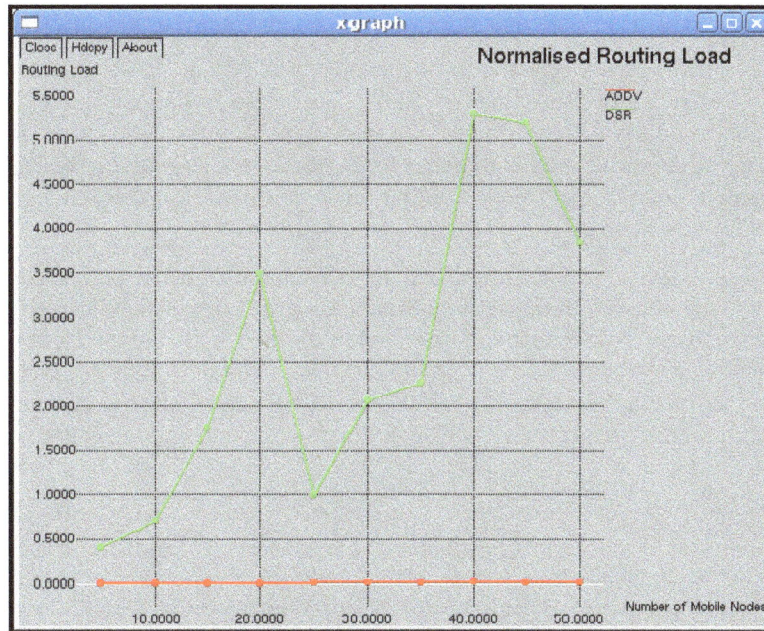

Figure 14. Normalized routing Load for AODV and DSR with
varying no. of Mobile Nodes

In the AODV routing protocol, when the number of nodes increases, initially throughput increases due to availability of large number of routes but after a certain limit throughput becomes nearly stable as shown in Figure 15. DSR also gives the consistent throughput but slightly smaller than AODV.

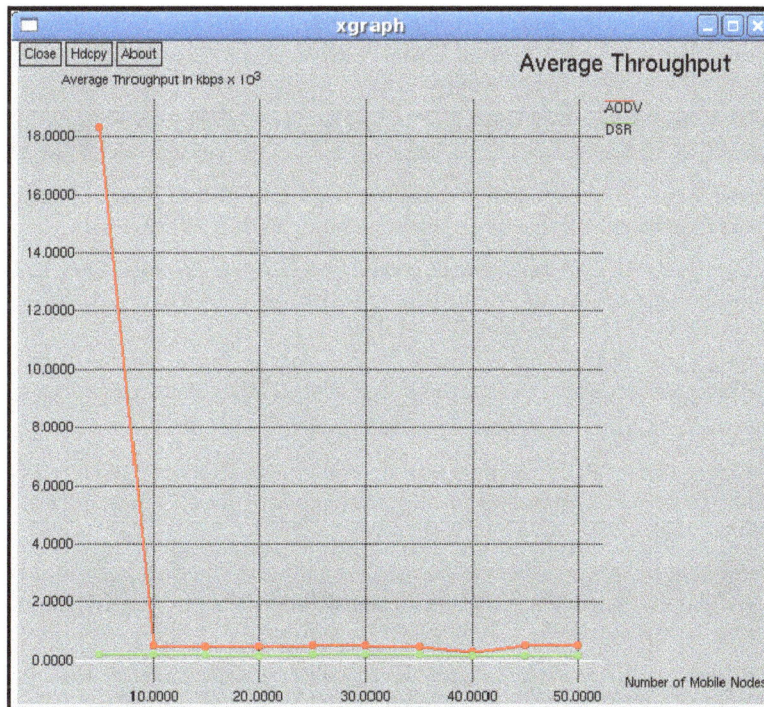

Figure 15. Throughput for AODV and DSR with
varying no. of Mobile Nodes

7. CONCLUSION

In this our simulation work, the routing protocols: AODV and DSR are evaluated for the application oriented performance metrics like packet delivery fraction, average end-to-end delay, throughput and normalized routing load with increasing the ten number of mobile nodes up to 50. As we increase the number of nodes for performing the simulation of AODV and DSR routing protocols, number of sent, routing and delivered packets changes, hence the performance parameters changes.

As a result of our studies, we concluded that AODV exhibits a better performance in terms of packet delivery fraction and throughput with increasing number of mobile nodes due to its on demand characteristics to determine the freshness of the routes. It is proved that the AODV has slightly higher average end-to-end delay than DSR. Our result also indicates that as the number of nodes in the network increases AODV and DSR gives nearly constant throughput. Considering the overall performance, AODV performs well with varying network size.

REFERENCES

[1]   J. Broch, D. A. Maltz, D. B. Johnson et.al.", A Performance Comparison of Multi-Hop Wireless Network Routing Protocols," Proceedings of the Fourth Annual ACM/IEEE International Conference on Mobile Computing and Networking (MobiCom'98), October 25-30, 1998, USA,pp.25-30.

[2]   S. R. Das, C. E. Perkins, and E. M. Royer, "Performance Comparison of Two On-Demand Routing Protocols for Ad Hoc Networks ", IEEE Personal Communications Magazine, Vol. 8, No. 1, February 2001, pp.16-29.

[3]   David Oliver Jorg, "Performance Comparison of MANET Routing Protocols In Different Network Sizes", Computer Networks & Distributed Systems, University of Berne, Switzerland, 2003.

[4]   K U Khan, R U Zaman and A. Venugopal Reddy,"Performance Comparison of On-Demand and Table Driven Ad HocRouting Protocols using NCTUns", Tenth International Conference on Computer Modeling and Simulation, 2008.

[5]   Saurabh Gupta "Analysis of simulation of ad hoc on demand distance vector routing protocol", National Conference on Advanced Computing and Communication Technology ACCT-10.

[6]   S. P. Setty et. al., "Performance evaluation of aodv in different environments", International Journal of Engineering Science and Technology Vol. 2(7), 2010, 2976-2981

[7]   Azzedine Boukerche, "A Performance comparison of routing protocols for Ad Hoc Networks", Parallel Simulations and Distributed Systems Research Laboratory, University of North of Texas,0-7695-0990-8/01/$10.00 (C) 2001 IEEE.

[8]   Arun Kumar B. R., Lokanatha C. Reddy, Prakash.S.Hiremath, "Performance Comparison of Wireless Mobile Ad hoc Network Routing Protocols" IJCSNS International Journal of Computer Science and Network Security VOL.8 No.6, June 2008.

[9]   S. Gowrishankar, et.al., "Scenario based Performance Analysis of AODV and OLSR in Mobile Ad Hoc Networks", Proceedings of the 24th South East Asia Regional Computer Conference, November 18-19, 2007, Bangkok, Thailand.

[10]   N Vetrivelan, Dr. A V Reddy, " Performance Analysis of Three Routing Protocols for Varying MANET Size", Proceeding of the International MultiConference of Engineers and Computer Scientists 2008 Vol II, IMECS 2008,19-21, Hong Kong.

[11]   Abdul Hadi Abd Rahman and Zuriati Ahmad Zukarnain, "Performance Comparison of AODV, DSDV and I-DSDV Routing Protocols in Mobile Ad Hoc networks", European Journal of Scientific Research, Vol.31 No.4 (2009), pp.566-576.

[12]   E. M. Royer, and C.K. Toh, "A Review of Current Routing Protocols for Ad Hoc Mobile Wireless Networks", IEEE Personal Communications, Vol. 6, Issue 2, pp. 46-55, April 1999.

[13]   NS -2, The ns Manual, Available at http: //www. isi.edu/nsnam/ns/doc.

[14]   C. E. Perkins, E. M. Royer, and S. R. Das, "Ad Hoc On- Demand Distance Vector (AODV) Routing", Internet Draft, draft-ietf- manet- aodv-10.txt, work in progress, 2002.

[15]   Satya Ranjan Rath, "Study of performance of routing protocols for mobile Adhoc networking in NS-2", NIT, Rourkela,2009.

[16]   David B. Johnson and David A. Maltz, "Dynamic Source Routing in Ad Hoc Wireless Networks", Computer Science Department, Carnegie Mellon University, Avenue Pittsburgh, PA 15213-3891.

[17]   Parma Nand and Dr. S. C. Sharma, "Performance study of Broadcast based Mobile Adhoc Routing Protocols AODV, DSR and DYMO," International Journal of Security and its Applications, Vol. 5 No. 1, January, 2011.

[18]   Rashmika N Patel," An Analysis On Performance Evaluation Of DSR In Various Placement Environments", Proceedings of the International Joint Journal Conference on Engineering and Technology (IJJCET 2010)

[19]   Anuj K. Gupta, Dr. Harsh Sadawarti and Dr. Anil K. Verma, "Performance analysis of AODV, DSR & TORA Routing Protocols," IACSIT International Journal of Engineering and Technology, Vol.2, No.2, April 2010 ISSN: 1793-8236.

[20]   Santosh Kumar,S C Sharma, Bhupendra Suman, "Simulation Based Performance Analysis of Routing Protocols Using Random Waypoint Mobility Model in Mobile Ad Hoc Network", Global Journal of Computer Science and Technology.

[21]   Rajesh Deshmukh, Asha Ambhaikar, "Performance Evaluation of AODV and DSR with Reference to Network Size", International Journal of Computer Applications (0975 – 8887) Volume 11– No.8, December 2010.

3

POWER CONTROL IN REACTIVE ROUTING PROTOCOL FOR MOBILE AD HOC NETWORK

Maher HENI[1] and Ridha BOUALLEGUE[2]

[1,2]Innovation of COMmunicant and COoperative Mobiles Laboratory, INNOV'COM
Sup'COM, Higher School of Communication, Ariana, Tunisia
[1] henimaher@gmail.com
[2] ridha.bouallegue@gmail.com

ABSTRACT

The aim of this work is to change the routing strategy of AODV protocol (Ad hoc On Demand Vector) in order to improve the energy consumption in mobile ad hoc networks (MANET). The purpose is to minimize the regular period of HELLO messages generated by the AODV protocol used for the research, development and maintenance of routes. This information is useful to have an idea about battery power levels of different network hosts. After storing this information, the node elect the shortest path following the classical model used this information to elect safest path (make a compromise) in terms of energy. Transmitter node does not select another node as its battery will be exhausted soon.

Any node of the network can have the same information's about the neighborhoods as well as other information about the energy level of the different terminal to avoid routing using a link that will be lost due to an exhausted battery of a node in this link.

Analytical study and simulations by Jist/SWANS have been conducted to note that no divergence relatively to the classical AODV, a node can have this type of information that improves the energy efficiency in ad hoc networks.

KEYWORDS

Ad-hoc Network, Routing protocol, energy consumption, AODV routing protocol, performance evaluation

1. INTRODUCTION

A Mobile Ad-hoc Network (MANETs) [1] is a collection of autonomous nodes or terminals that communicate together by forming a multi-hop radio network and maintaining connectivity decentralized. The nodes can move and their network topology may be temporal. Each node acts as a customer, server and router. In such network, there is no centralized administration. Each node can join the network or it leave at any time.

Routing protocols in such networks can be classified mainly into three categories:

• Proactive routing protocols: They are based on the same principle as wired networks routing. Paths in this type of routing are calculated in advance. Each node maintains multiple routing tables by exchanging control packets between neighbors. Indeed, if a node wants to communicate with one another, it has the ability to view local routing table and create path it needs. OLSR [2](Optimized Link State Routing) and FSR [3](Fisheye State Routing) are examples of proactive routing protocols.

- Reactive routing protocols: On the Contrary of proactive protocols, reactive protocols calculate the route on request. If a source node needs to send a message to a destination node, then it sends a request to all members of the network. After receiving the request, the destination node sends a response back to the source. However, the routing application generates a slow pace because of the research paths which can degrade application performance. Such protocol has the disadvantage of being very costly in terms of energy and packets transmission when determining routes but has the advantage of not having to hold unused information in routing tables. AODV [4] is an example of reactive protocols which are described below.

- Hybrid routing protocols: Hybrid routing protocols or "mixed" combine the previous two types of routing (proactive and reactive). The proactive protocol is applied in a small area around the source (limited number of neighbors), while the reactive protocol is applied beyond this perimeter (distant neighbors). This combination is performed in order to exploit the advantages of each method and overcome their limitations. ZRP [8] (Zone Routing Protocol) and CBRP [7] (Cluster Based Routing Protocol) are two major examples of hybrid protocols.

One of the major and most critical factors in ad hoc networks is the limited battery energy. A large amount of works is focused on this setting to reduce the consumption of batteries. The waste of energy may be due to the regular exchange of unnecessary control messages to have more reliability.

AODV (Ad hoc On-Demand Distance Vector Routing Protocol) is a reactive routing protocol designed by Charles E. Perkins and Elizabeth M. Royer [4]. This protocol uses four types of control messages in the aim to send data packets. The first type is HELLO messages. This type of messages, exchanged periodically to maintain a neighborhood base. RREQ, RREP, RRER, are used to establish a path to destination when any node wants to send a data. This number of control packet has a signified effect of the waste of resources.

To overcome the problem of energy consumption in this protocol, we designed a new solution that reduces the HELLO messages number exchanged and to include the factor of energy consumption that will be useful later for the routing messages. Firstly, we minimize the exchange number of Hello messages. Secondly, we replace the regular periodic instant of sending hello message by another proportionally to the energy stored in the battery of the node. The node receiver of this hello message, do the inversely action to extract information proportionally to the node sender energy, and the same information enclosed in hello message.

Insert this parameter does not affect the operation or the information included in messages exchanged and then we can obtain new information that we can use to elect path. We call the new protocol PC-AODV (Power Control AODV).

This paper is organized as follows: we present the AODV routing protocol in Section 2. In Section 3, we detail the related work, we expose the used model and parameters in section 4, and we formalize our solution and we present the new protocol called PC-AODV in section 5. In the rest of the paper, we illustrate, in section 6, an analytical comparative study between the classical and the new protocol. In section 7, we present a simulation evaluation, of the tow protocol using JiST/SWANS simulator. We conclude this paper and present future work in Section 8.

2. AD HOC ON-DEMAND DISTANCE VECTOR ROUTING PROTOCOL

2.1. Overview

AODV [4] [5] is a reactive protocol that is based on the concept of distance vector routing protocols as its name mean. The algorithm of AODV is inspired from the combination of a proactive and a reactive protocol [24, 25]. The path discovers and maintains is similar to the process used in DSR [26]. The uses of HELLO message exchange to establish a neighborhood base and sequence number method are used in DSDV [27]. AODV present more performance in static and bulky networks. These factors present major challenges to MANETs routing protocol researchers.

AODV performs route discovery request and saves only used routes in the routing table. It use four different control message called HELLO, RREQ, RREP, and RRER message. In order to transmit data packets, it broadcasts a route request RREQ (Route REQuest message) in the wholly networks. Three cases are possible upon receipt of a RREQ message by any node. In the first, if the node that received this message provides a route to the requested destination in its routing table, it responds with another type of message RREP (Route REPly message). In the second case, if it hasn't information about the destination, it will retransmit the message to its neighbors that have not yet received. If all the neighbors have received the same message and/or the node has lost the connection, it responds with an error message RERR (Rout ERRor message). After receiving a reply message, the source node starts sending data packets along the shortest path.

Other than these messages, AODV uses only one type of periodic message is HELLO message, in order to maintain the Neighborhood basis.

In either case, the source node waits for a predefined timeout, the route establishment response to the destination, and then it retransmits another RREQ by increasing the maximum number of hops (TTL: Time To Live). If after repeating this process a limited number and the source get nothing, it declares the absence of this destination.

To maintain routes, AODV use an ACTIVE_ROUTE_TIMEOUT (ART) that equal to 3 second [28]. If and defined routes between tow nodes, is not used within this period, then this node is not sure if this route is yet available or not, it rebroadcast a RREQ if needs

2.2. Motivation

To exchange these types of route establishment messages, each node periodically exchange HELLO messages to maintain a neighborhood base and the routing table. Since the regular exchange of both control messages amplifies considerably the energy consumption, and bandwidth.

Generally, MANETs are characterized by limited energy and bandwidth. With the exchange of this considerable number of control messages to establishment of routes, this aggravates the resources and performances, precisely in the case of bulky networks. One of major causes of exchange of these message is the lost of paths, result of the exhausting of a node battery. To overcome this problem, we suppose that all nodes composed the networks have information about the energy stored in the batteries of its neighborhood. It avoid the routing using a node that it battery will be exhausted and take into account the nodes that can be used.

To do this, we propose a mechanism that reduces the exchange of this type of message and use it to inform about the energy state of the other nodes in keeping the same performances of the standard protocol. Using hello message, all nodes exchange a new type of information about the energy stored in battery, without changing the fields composed this message, and simply tuning the instant of sending it.

3. RELATED WORK

The work done in this context could be grouped into two major groups; the first describes methods for reducing energy consumption in the AODV protocol with diversifying the routing strategy, and the second present's methods to reduce numbers of control messages in order to reduce the cost of consumption of energy.

In [14] authors propose a new version of AODV called (MAODV) derived from the AODV routing protocol by considering the bit error rate (BER) at the end of a multi-hop path as the metric to be minimized for route selection. In [15], authors integrated the transmit power control and load balancing approach as a mechanism to improve the performance of on-demand routing with energy efficiency. M.Veerayya, V. Sharma and A. Karandikar propose in [16] a cross-layering approach to exchange information about the residual energy in nodes to perform quality of service. In [17] a new mechanism is proposed to set a timeout for a path. A path considered broken if a node leave by following the exhaustion of its energy. In [18] authors integrate the runtime battery capacity in routing protocol and the estimated real propagation power loss, obtained from sensing the received signal power. This solution is independent of location information and using the propagation, they estimate the energy loosed. Another type of the proposed work which aims to reduce the overhead of AODV to achieve energy efficiency, as described in [19]. Authors propose a new method in order to reduce overhead in AODV in urban area by predicting links availability. By predicting neighbor nodes positions it can be determined probability of link failure.

In [20] S.B. Kawish, B. Aslam, S. A. Khan studies the behavior of AODV in a fixed networks and those exhibiting low mobility with a view to highlight the reasons for reducing overhead and then reduce the energy consumption. The same authors present in [21] an improvement in their idea of using route timeout adjusted to reduce the overhead.

In [22] Authors propose a new version of AODV an on-demand routing algorithm based on cross-layer power control termed as called CPC-AODV (Cross-layer Power Control Ad hoc On-demand Distance Vector) taking account of the geographic location of nodes, the energy of packet transmission. Furthermore, the approach presented in [23] consists of an algorithm that enables packet forwarding misbehavior and Loss Reduction based detection through the principle of conservation of flow on the routing protocol group nodes.

First, unlikable the other proposed solution, our protocols, does not minimize the number of messages or the overhead, or use geographic coordinates of the nodes or the channel access using the MAC layer. Our solution simply changes the periodicity by random time for the receiver and set by the power level of the node battery the transmitter. This is an important feature and has a profound effect on energy consumption which could sustain the behavior of protocol. It is an available approach to incorporate routing protocols with power control in ad hoc networks.

4. USED MODEL

We use a network composed by four nodes (node A, B, C and D) with bidirectional or symmetric links between them. The communication range is circular with a diameter of 250 meters.

Table 1. Used Variables.

Variable	Designation
E_x	Energy stored in the node x battery
E_R	Resultant energy
K_x	1/Ex
HELLO(x)	Message HELLO
$H_{ACK}(x)$	Hello message acknowledgment
T_{ACK}	reception time of HELLO message acknowledgments
Δt	acknowledgment period
HI	HELLO_INTERVAL
Nn	Node's neighborhood number nodes

Our goal is primarily to have an idea about the quantity of energy stored in batteries for neighbors. The parameters used to define the model are defined in Table 1. The topology used is shown in Figure 1.

In our model, we decrease the number of HELLO messages by increasing the time between two messages. Assuming that the period between two successive HELLO messages is proportional to the neighbors number according to the equation (1).

$$HI_{PC-AODV} = N_n * HI_{AODV} \qquad (1)$$

We will reduce this interval to receive Acknowledgment that have the same content as the hello messages, but which allows to know the battery level of the other nodes.

Ki is inversely proportional to the battery's energy. Assuming that the máximum level of the battery power is 15Kw [29], in our example Kc is the node C factor, then the level of it battry power is

$$E_c = \frac{1}{K_c} * E_{Max} = \frac{1}{6} * 15Kw = 2,5Kw$$

Respectively

$$E_B = 5Kw, E_A = 7,5Kw, E_D = 3,75Kw$$

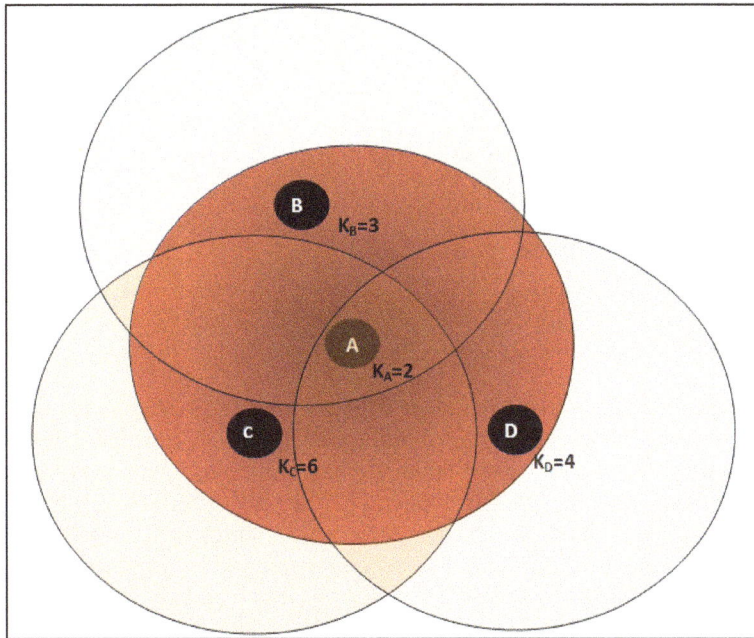

Figure 1. The neighbourhood topology used for the description

After receiving a hello message from node A, all node start send a Hello acknowledgments at an instant proportionally to the factor K. then, Node B will send, the first, an acknowledgments to A, node D and then node C (see Figure 2). Using this concept, node A do the inversely process to extract the level of neighbour's battery.

5. PC-AODV CONCEPT

Our solution is illustrated in Figure 2. After sending a Hello message, the node A starts receiving acknowledgments from its neighbors. The parameter δt is assumed known by all nodes and is defined in the HELLO message. We chose a parameter K the inverse of the energy stored in order to have the first acknowledgment of the node that has the maximum energy, and either receive or not the final acknowledgment at the end of the period. If the energy of a node is negligible, for example, you will not receive acknowledgment during the period T_{ACK}.

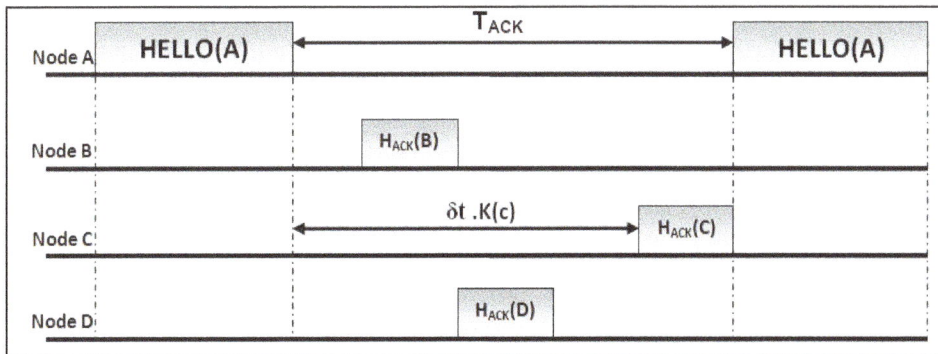

Figure 2. HELLO transmission and reception model

To better understand the phenomenon, we take the example of figure 1where K (B)=3, K(C) = 6 and K(D) = 4.where

$$K(i) = \frac{1}{node\_i\_Battery's\_level}$$

We assume that node A select $\delta t = 2$.

Thus the sent start time to of these nodes are:

$t_C = \delta t \times K(C) = 2 \times 6 = 12$ Time_unity.

$t_B = \delta t \times K(B) = 2 \times 3 = 6$ Time_unity.

$t_C = \delta t \times K(D) = 2 \times 4 = 8$ Time_unity.

So the first node who starts sending is B , the next one is D , then node C.

After receiving various Hello acknowledgments node A registers and updates the localization information and battery energy in the neighborhood base. An Acknowledgement of HELLO message sent by a neighbor B and is itself another HELLO message generated by this node and containing the same information as classical message. The only difference is the non-periodicity of this message in order to have the desired information. Information about energy can be further used for the dissemination of other control messages or data. In the event that a node in the neighborhood has a battery exhausted, the node A avoids the flow of messages using this node.

After collecting information concerning the batteries level, a node source then chooses more than the shortest path, the safest path. Indeed, after selecting of shorter path,and sending a data message, a node can be left this way and so the link considered broken and then we have loss of message. However, in the case of safe path the source nodes can elect the path that contain no node with an exhausted battery.

6. ANALYTICAL COMPARATIVE STUDY

In this section, we present a comparison between AODV and PC-AODV. We compare analytically the two concepts in the same condition and parameters.

To prove the energy efficiency protocol for PC-AODV, we consider, as a metric, the energy required for the correct transmission of a packet from mobile node i to node j. On the first hand we take the energy Eij formula presented in [9] and [10].

$$E_{i,j} = \frac{MP_i}{RP_c(\delta_{i,j})} \tag{2}$$

Where M denotes the length of the packet, Pi is the transmission power of node i, R represents the data transmission rate and Pc(δij) is the probability of correct reception of a packet from node i to node j, with δij equal to the SIR (Signal to Interference Ratio) [30] of link (i,j). This expression depends on the transmission quality and characteristics of link between the two nodes.

On the other hand, we take the second formula developed in [11]

$$E(p,n_i) = P(p,n_i) * t_p = i * v * t_p \quad (3)$$

Where P(p,ni) is the power to transmit a packet p by any node ni, v is the node battery voltage and, i is the current (in Ampere).

i *v is the constant party of the equation and dependent on the technical characteristic of the host. Using [12], the time taken to transmit the packet p (in sec) is presented by (2):

$$t_p = \frac{p_h}{6*10^6} + \frac{p_d}{54*10^6} \quad (4)$$

ph is the number of bits of the packet header, and pd is the payload presented in [17] by the following formula:

$$p_d = 256(228(data) + 8(UDP) + 20(IP)) \quad (5)$$

In order to compare the two concepts, we consider two nodes x and y and assume that the parameters for message transmission are the same in both cases. Indeed, we take both equations (2) and (3) representing the energy, if we assume that the conditions are the same (same battery level, even message length in equation (3), the same transmission condition, even, message length for the equation (2)). Thus, we consider that the energy transmission of a HELLO or an acknowledgment message from source x to destination y in our solution equal to an accurate value C (equation (6)), since both have the same length and sent in the same condition.

$$E_{x,y} = C \quad (6)$$

According to [9] and [10] the resulting energy ER (Resultant Energy) required for the routing of messages over a period t is determined by

$$E_R = \sum_{t,(\forall(i,j)\in N)} E_{i,j} \quad (7)$$

N, is neighborhood number of a node. So if we consider the same network in Figure 1, we present by equation (8) the energy consumed by the four nodes in a period TSIM = 60 seconds with Hello Interval (AODV)(HI (AODV)) is equal to1 second according to the draft:

$$E_{total}(AODV) = \frac{4(nodes) * C * T_{SIM}}{HI(AODV)} \quad (8)$$

In the case of PC-AODV and since almost the entire node performing the same transmission the resultant energy is given by:

$$E_{total}(PC - AODV) = E_R(A) + E_R(B) + E_R(C) + E_R(D)$$

$$= E_{HELLO} + 3 * E_{ACK} = \frac{4(nodes) * C * T_{SIM}}{HI(AODV)} \quad (9)$$

Proof: let take HI (PC_AODV) = number of node responding with acknowledgment multiplied by HI(AODV). The node A that transmits a first hello message take the same interval as AODV protocol.

$$E_{HELLO} = \frac{C * T_{SIM}}{HI(AODV)}$$

$$E_{ACK} = \frac{3(nodes) * C * T_{SIM}}{HI(PC - AODV)} = \frac{3(nodes) * C * T_{SIM}}{3 * HI(AODV)} = \frac{C * T_{SIM}}{HI(AODV)}$$

According to equation (8) and (9) we note that the energy consumption due to the number of HELLO messages (adding the acknowledgment in the case of PC-AODV) is the same in both versions. So we do not degrade the classic version in terms of energy. Otherwise, our solution provides information about energy stored by the neighborhood under the same conditions.

7. SIMULATION RESULT

After the analytical comparison of the two protocols, we compare the number of messages exchanged using the simulators and the JIST / SWANS [13]. Simulations of these two protocols were made on the same simulations model (same parameters values and even traffic patterns).

The platform JIST/SWANS is a high-performance discrete event simulator, developed in Cornell University. JIST (Java in Simulation Time) is a driving simulation of discrete event which runs over a Java virtual machine (JVM). SWANS (Scalable Wireless Ad-hoc Network Simulator) are a network simulator that runs on top of JIST.

Through this simulator, we analyzed the overhead metric of the routing ad hoc protocol AODV and PC-AODV.

The Figure 3 shows the number of HELLO messages sent and received in case of AODV and on the same figure we also represented the sum of HELLO and HELLO_ACK messages sent and received in case of PC-AODV. Both of simulations are made in a period of 1800 seconds, with a variable number of nodes, static and randomly distributed nodes over an area of 1000x1000 meters. In Figure 4, 5 and 6 we have represented respectively Route Request, Route Reply and Route Error messages exchanged used by AODV and PC-AODV for route establishments. We also presented in figure 7 the routing overhead of AODV and PS-AODV.

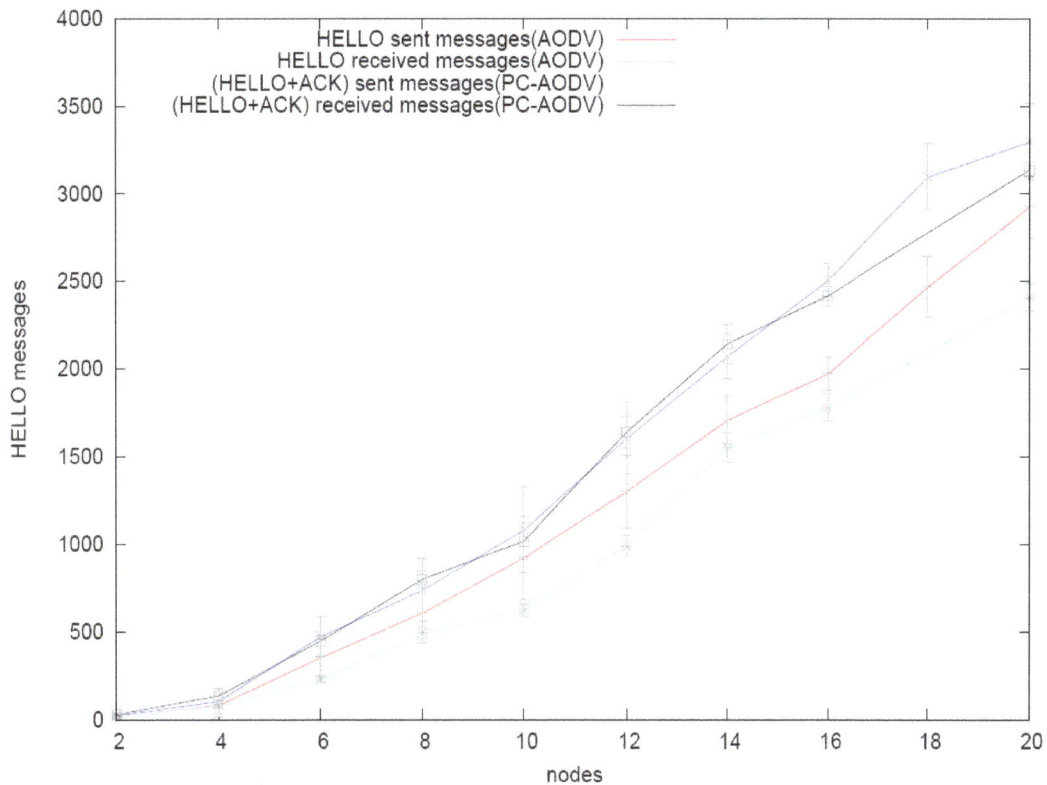

Figure 3. Exchanged Hello and Hello_ ACK messages

Through figure 3 we found that the number of HELLO messages in the case of AODV increases with the number of the node approximately linearly. Since the HELLO messages are periodic, so this increase depends of, the number of nodes, the period of HELLO messages, and the duration of simulation. Otherwise, the curve of HELLO messages reception is above than the curve of transmission and is solved by the fact that a hello message sent by one node will be received by the n neighbors of this node, and then a message is sent once and received more than once.

On the amounts of HELLO_ACK messages acknowledgment is slightly lower than the number of HELLO message, as we noted earlier, the nodes that has a very low energy cannot send acknowledgments in the period for reception of this type of messages. We used an energy model to assign a random power level for each node between two maximum and minimum thresholds, in order to have closely conditions of real network.

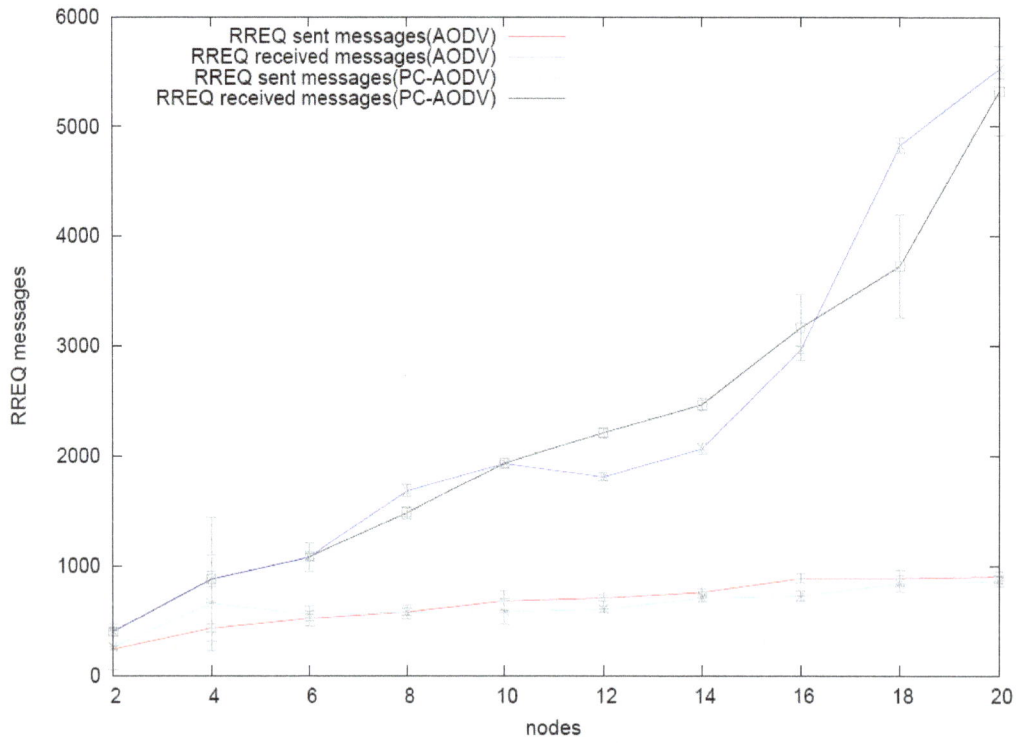

Figure 4. RouteREQuest Message exchanged

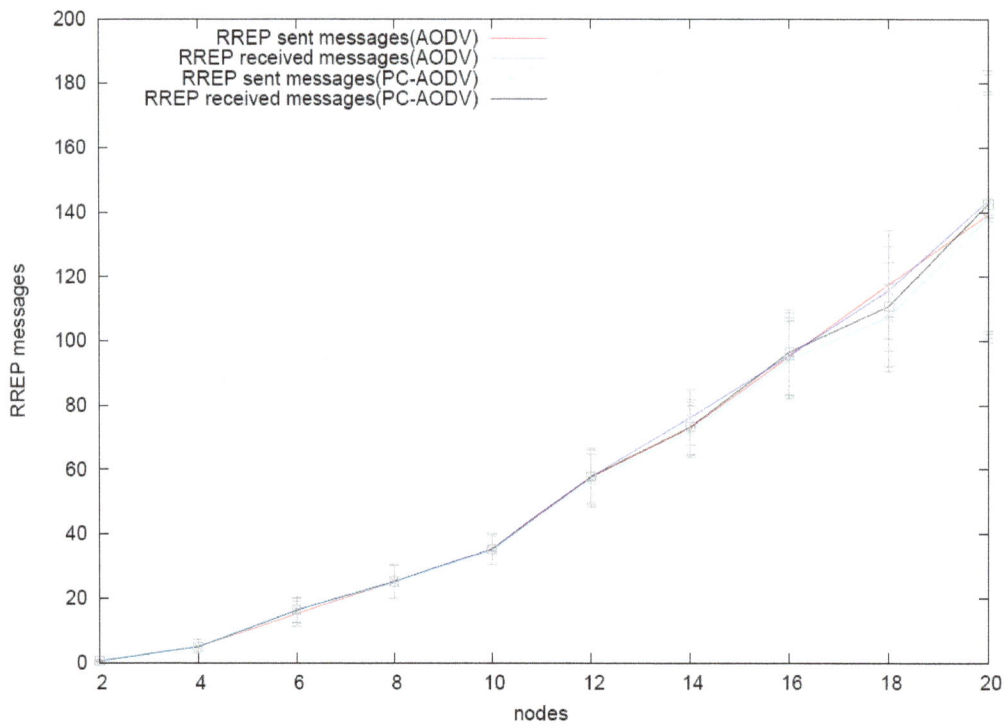

Figure 5. Exchanged Route REPlay messages

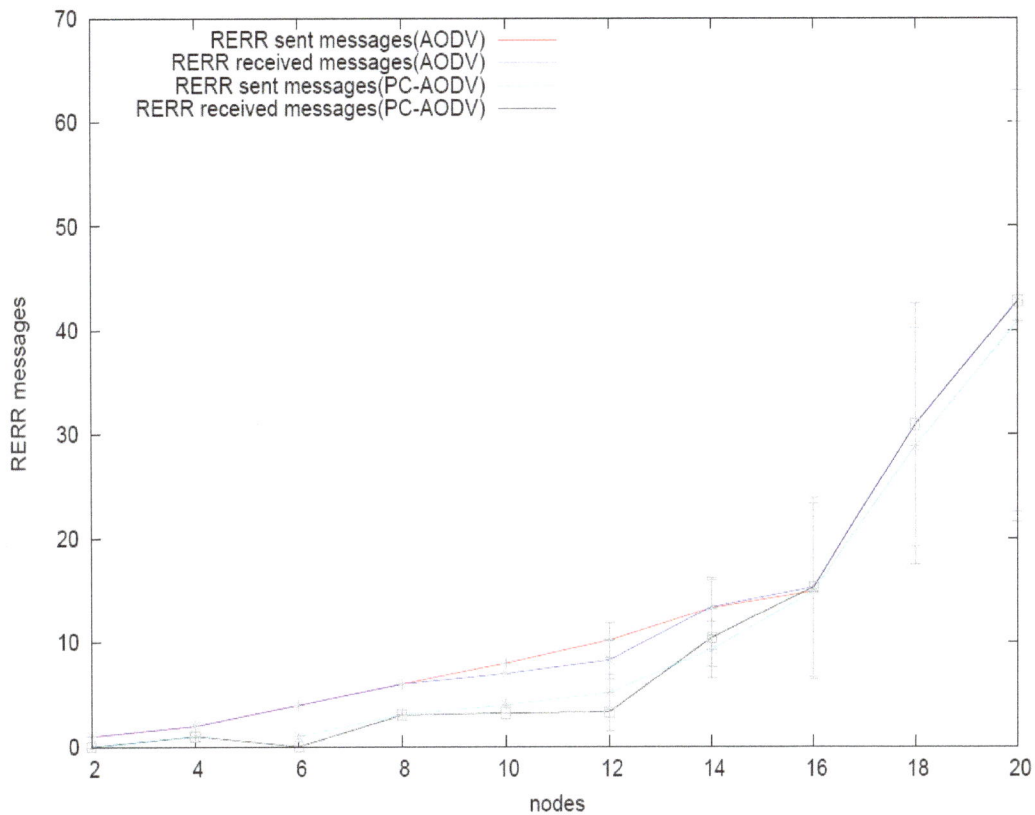

Figure 6. Exchanged Route ERRor messages

Figures 4, 5 and 6 show the influence of our solution on the number of control (other than Hello messages) exchanged. These two curves show that our solution allows AODV to generate the same traffic control and data but also shows that gains information about the energy stored in the batteries of the nodes.

So we note that the curves coincide with a slight difference due to the condition of random simulation to approximate a real network. The number of messages generated shows the overhead of network to establish paths and to send data packets. We can deduce two main metric, packet delivery ratio and overhead [31],[32],[33].

The letter is the most metric needed in Ad-Hoc network simulations. We found several definitions of this metric according to the parameters set of simulations. We chose one that measures the overhead of network control messages. It is equal to the number of control messages broadcast in the network divided by the sum of this number and the number of data messages sent. Overhead presented in Figure 7.

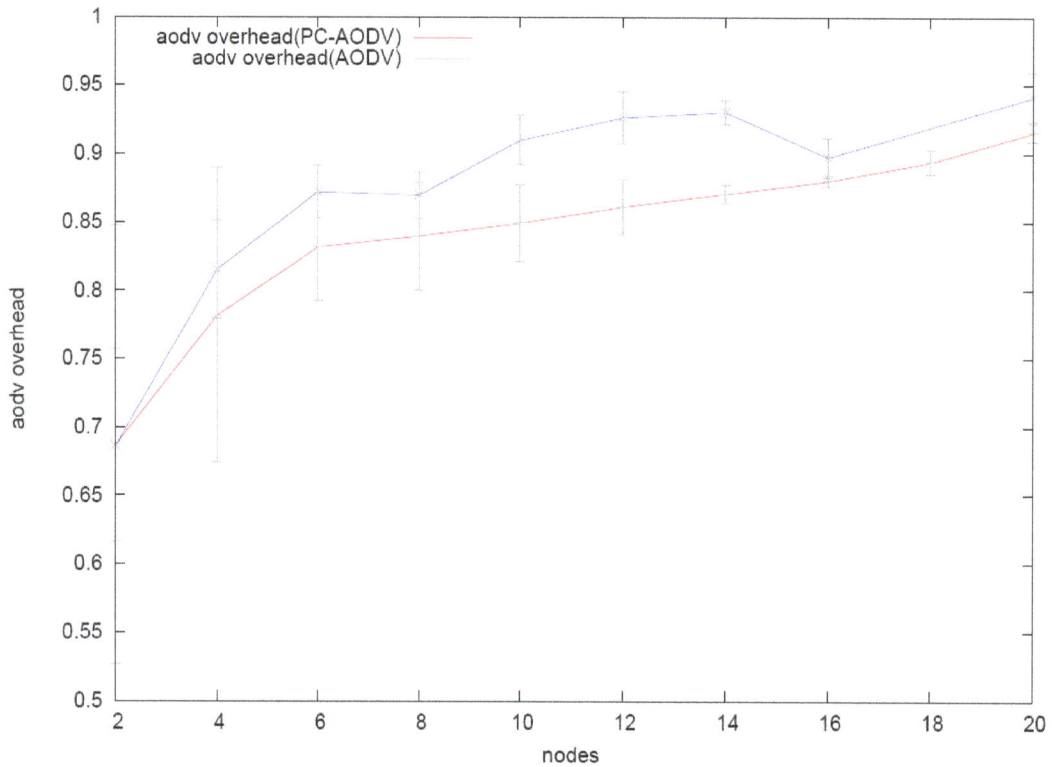

Figure 7. Routing overhead

Through this curve we first note that as the number of nodes increases, the network become charged, and it shows the disadvantage that AODV is not appropriate to the dense network. In addition the overhead of PC-AODV is lower than AODV, that because the number of HELLO message is smaller, so there was a slight decrease compared to AODV.

The packet delivery ratio is another mainly metric, it represent the safety of reception through the routing protocol used. It equal to number of received messages divided by the number of transmitted messages. We used the RREP messages, since this type of message follows the shortest path found (calculated using Dijekstra algorithm []) after the broadcast of RREQ messages. Packets delivery ratio is shown in Figure 8 using the equations (10):

$$PDR = \sum_{Transmission} \frac{\mathrm{Re}\,ceived\_Messages}{Transmitted\_Messages} \quad (10)$$

This solution provides to nodes to look more than shortest path, the safe path (the path that contains enough energy to routing packets). Indeed, using the batteries level information, the node doesn't choose a path that contain a node that risk to leave the network because it battery is empty. In the classical case, sometimes there was a link failure caused by the departure of a node due to the depletion of its battery, and this will reset again another route and that path we broadcast many more route discovery message.

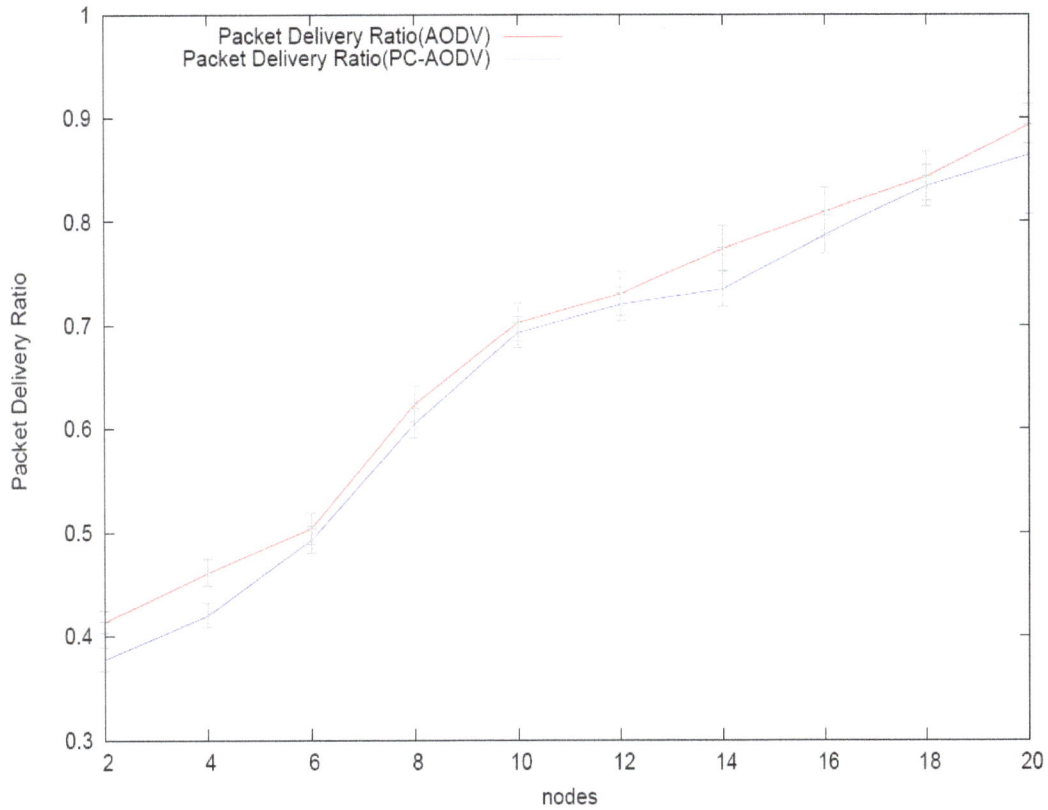

Figure 9. Packet Delivery Ratio

8. CONCLUSION:

In this paper we presented a new solution for the exchange of HELLO messages in AODV routing protocol. We have shown that our solution can provide knowledge about the levels of stored energy of the nodes constituting the network without affecting the operation of the protocol. After saving this new information, a node given in the network can choose the shortest path that contains enough energy for the correct routing of data packets, thus winning in terms of the ratio of the packets.

In future work, we will evaluate the PC-AODV performances in different topologies and different types of mobility to demonstrate the robustness of this Protocol and the benefits provided.

We will also set parameter that we guarantee the good receptions of acknowledgment, we will calculate the collision probability of acknowledgments to find factor to ensure the receipt of such messages, and therefore a hello sender node receives the acknowledgments messages in highly accurate moments.

REFERENCES

[1]   S. Corson, J. Macker, "Mobile Ad hoc Networking (MANET):Routing Protocol Performance." Issues and Evaluation Considerations, http://tools.ietf.org/html/rfc2501, 1999.

[2]   T. Clausen, P. Jacquet, "Optimized Link State Routing Protocol (OLSR)", Draft IETF RFC 3626, http://tools.ietf.org/id/draft-ietf-manet-olsr-11.txt, 2003.

[3]   G.Pei, M.Gerla, T.W.Chen, "Fisheye State Routing: A Routing Scheme for Ad Hoc Wireless Networks", Proceedings of the IEEE International Conference on Communication (ICC), Pages 70-74, Juin 2000.

[4]   C. Perkins, E. Belding-Royer, "Ad hoc On Demand Distance Vector Routing", the 2nd IEEE Workshop on Mobile Computing Systems and Applications (WMCSA'99), New Orleans, Louisiana, USA, 1999, pp 25-26.

[5]   C. E. Perkins, E. M. Belding-Royer, S. R. Das, "Ad hoc On-Demand Distance Vector (AODV) Routing", Draft IETF,http://tools.ietf.org/html/draft-ietf-manet-aodv-13, 2003.

[6]   C.E. Perkins, S.-J.Lee, E.M. Belding-Royer, "Scalability study of the ad-hoc on-demand distance vector routing protocol". Int. J. Netw. Manag., 13(2), 2003.

[7]   M. Jiang, J. Li, Y.C. Tay, "Cluster Based Routing (CBRP)", Draft IETF, http://tools.ietf.org/html/draft-ietf-manet-cbrp-spec-01, 1999.

[8]   Z. J. Haas, M. R. Pearlman, P. Samar, "The Zone Routing Protocol (ZRP) for Ad Hoc Networks", Draft IETF, http://tools.ietf.org/id/draft-ietf-manet-zone-zrp-04.txt, 2003.

[9]   D.J. Goodman, N.B. Mandayam, "Power control for wireless data", IEEE Personal Communications Magazine 7(2) , 2000.

[10]   N. Nie, C.Cristina, "Energy efficient AODV routing in CDMA ad hoc networks using beamforming", Journal on wireless Communications and Networking EURASIP, Volume 2006, Issue 2, 2006.

[11]   L. M. Feeney, M. Nilsson, "Investigating the energy consumption of a wireless network interface in an ad hoc networking environment", INFOCOM, 2001.

[12]   G.David,M.Narayan, "Power Control for Wireless Data," IEEE Personal Communications , April 2000.

[13]   R. Barr, Z. J. Haas, JiST/SWANS. http://jist.ece.cornell.edu, 2005.

[14]   G. Ferrari, S. A. Malvassori, M. Bragalini, O. K. Tonguz, Physical Layer-Constrained Routing in Ad-hoc Wireless Networks: A Modi_ed AODV Protocol with Power Control, IWWAN, 2005.

[15]   M. Tamilarasi, T.G. Palanivelu, Integrated Energy-Aware Mechanism for MANETs using On-demand Routing , International Journal of Computer and Information Engineering,2008

[16]   M. Veerayya, V. Sharma, A. Karandikar, SQ-AODV: A novel energy aware stability based routing protocol for enhanced QOS in wireless ad-hoc networks, MILCOM 2008

[17]   M. Tamilarasi, T.G. Palanivelu, Adaptive link timeout with energy aware mechanism for on-demand routing in MANETs, Ubiquitous Computing and Communication Journal, 2010.

[18]   R.C.Shah,J.M. Rabaey, Energy aware routing for low energy ad hoc sensor networks, WCN, 2002

[19]   R. Ghanbarzadeh, M. R. Meybodi, Reducing message overhead of AODV routing protocol in urban area by using link availability prediction, Second International Conference on Computer Research and Development, 2010.

[20]   S.B. Kawish, B. Aslam, S. A. Khan, Reducing the Overhead Cost in Fixed and Low Mobility AODV Based MANETs, Proceedings of the International Multiconference on Computer Science and Information Technology,2006.

[21]   S.B. Kawish, B. Aslam, S. A. Khan, Reduction of Overheads with Dynamic Caching in Fixed AODV based MANETs, World Academy of Science, Engineering and Technology,2006.

[22]   H. Huang, G. Hu, F. Yu, A Routing Algorithm Based on Cross-layer Power Control in Wireless Ad Hoc Networks, Communications and Networking in China (CHINACOM), 2010 .

[23]   K.T. Sikamani, P.K. Kumaresan, M. Kannan, R. Madhusudhanan, Simple Packet Forwarding and Loss Reduction for Improving Energy Efficient Routing Protocols in Mobile Ad-Hoc Networks, European Journal of Scientific Research, 2009.

[24]   S. R Das, C. E Parkins and E. M Royer: Performance Comparison of Two on Demand Routing Protocols for ADHOC Networks, in Proceedings of IEEE INFOCOM 2000, 2000.

[25]   M. Cagalj, Performance Evaluation of AODV Routing Protocol: RealLife Measurements, SCC, 2003.

[26]   J.broch, David B.johnson and D. A.Maltz, The Dynamic Source Routing Protocol for Mobile Ad Hoc Networks, Draft IETF, Febrary 2002.

[27]   C.Perkins and P.Bhagwat, Highly Dynamic Destination-Sequenced Distance-Vector Routing (DSDV) for mobile Computrrs, SIGCOMM'94.

[28]   M. Lakshmi and P. E. Sankaranarayanan, Performance Analysis of Three Routing Protocols in Mobile Ad Hoc Wireless Networks, Asian Journal of Information Technology, 2005.

[29]   M.R. Jongerden and B.R. Haverkort, Battery Modeling, Design and Analysis of Communication Systems, 2008.

[30]   M. Ghimire, R. Al-Rizzo, H. Akl and R. Y. Chan , Channel assignment in an IEEE 802.11 WLAN based on Signal-To-Interference Ratio, Electrical and Computer Engineering, CCECE, 2008.

[31]   Y. Li and H. Man, Three load metrics for routing in ad hoc networks, Vehicular Technology Conference, 2004.

[32]   S.Wang, Z. Qiu, A link type aware routing metric for wireless mesh networks, ICSP'8 , Beijing,2006.

[33]   G.Karbaschi and A. Fladenmuller,A link-quality and congestion-aware cross layer metric for multi-hop wireless routing, IEEE International Conference on Mobile Adhoc and Sensor Systems Conference,Paris, 2005.

4

A Review and Comparison of Quality of Service Routing in Wireless Ad Hoc Networks

Sunita Prasad[1] and Zaheeruddin[2]

[1]Centre for Development of Advanced Computing, NOIDA, India
sunitaprasad@cdacnoida.in
[2]Department of Electrical Engineering, Jamia Millia Islamia, India
zaheer_2k@hotmail.com

ABSTRACT

Quality of Service (QoS) guarantees must be supported in a network that intends to carry real time and multimedia traffic. IETF RFC 2386 defines QoS as a set of service requirements to be met by the network while transporting a packet stream from source to the destination. The dynamic network topology and wireless bandwidth sharing makes QoS provisioning far more challenging in wireless networks as compared to the wired counterparts. The support for the QoS services is underpinned by QoS routing. A QoS routing protocol selects network routes with sufficient resources for the satisfaction of the requested QoS parameters. The goal of QoS routing is to satisfy the QoS requirements for each admitted connection, while achieving global efficiency in resource utilization. The problem of QoS routing with multiple additive constraints is known to be NP-hard. This requires the QoS dynamics to be fully understood before it can be implemented in wireless ad hoc networks. The paper discusses the issues involved in QoS routing and presents an overview and comparison of some existing QoS routing protocols. The article concludes with some open issues for further investigation.

KEYWORDS

Wireless ad Hoc Networks (WANET), Quality of Service (QoS), QoS Routing, Multi-Constrained Path (MCP) problem.

1. INTRODUCTION

A Wireless Ad hoc Network (WANET) consists of a collection of mobile nodes connected by wireless links which can be created on-the-fly without using any infrastructure or administrative support [1]. These networks are characterized by *self-organization and autonomy*. The wireless ad hoc network has been attracting increasing attention of researchers owing to its good performance and special application scenarios. For example, it can be used in military operations for fast deployment of troops in hostile and unknown environments, search and rescue operations for communication in areas having little or no wireless infrastructure support, disaster relief operation where the existing infrastructure is destroyed or left inoperable and commercial application like enabling communication in exhibitions, conferences and large gatherings. Examples of wireless ad hoc networks are Zigbee and Bluetooth networks. The perception that wireless ad hoc networks are simply a wired network with cables replaced with antennas is a common misconception. The unique characteristic of wireless ad hoc networks like dynamic topology and resource constraint distinguishes it from wired networks and necessitates the need of special solutions in these networks [2].

With the advancement of technology, the wireless and portable computers and devices are becoming more powerful and capable. There is a growing desire for the wireless networks to support real time multimedia applications. Such applications require the network to provide guarantees on the Quality of Service (QoS). QoS is the ability of the network to provide some level of assurance for consistent network data delivery [3]. The network is expected to guarantee a set of measurable prespecified service attributes to the users in terms of delay, delay variance (jitter), bandwidth, probability of packet loss, etc. Power consumption and service coverage area are the two other QoS attributes that are more specific to WANETs [14]. The goal of QoS is to provide some level of predictability and control in the network behavior and at the same time achieve global efficiency in resource utilization.

However, the current WANET architecture is not adequate for the support of more demanding applications. With Internet as the basic model, ad hoc networks have been initially considered only for best-effort services. QoS provisioning in wired networks is based on two approaches: (1) Overprovisioning and (2) Network Traffic Engineering. The first method consists of offering huge amount of resources to accommodate all the demanding applications. This method is difficult to implement in wired networks and completely infeasible in wireless networks where there is a scarcity of resources. The second method classifies the ongoing traffic in the network and processes them according to some set of rules with the joint goals of good user performance and efficient use of network resources. These solutions cannot be directly applied to wireless ad hoc networks because of their unique characteristics. Thus, in the last few years, QoS for ad hoc networks has emerged as an active area of research [8][13][16][17][18][21].

The support for QoS services in the network depends to a large extent on QoS routing. QoS routing is a routing mechanism under which paths that satisfy the QoS constraints are determined based on the knowledge of resource availability in the network and the QoS requirements of the flows. The objective of QoS routing is to find a feasible path between a source–destination pair, if one exists, that optimizes the use of network resources and satisfies the required QoS guarantees. It is evident that QoS routing is a constrained combinatorial optimization problem [12]. [4] presents an exhaustive review on QoS routing methodologies for ad hoc networks.   This paper presents an overview and comparison of some existing QoS routing protocols. We also discuss the future trends in providing QoS guarantees in the network. The paper is organized as follows- Section 2 reviews the QoS routing and challenges in providing QoS guarantees in wireless ad hoc networks. Section 3 presents the QoS metrics. Section 4 analyzes some of the QoS aware routing protocols. Finally the last section concludes the paper and gives some future direction.

2. QoS ROUTING

The support of QoS services is underpinned by QoS routing. Many routing protocols for wireless networks such as AODV [5] and DSR [6] use best effort routing where all nodes within the range compete for the shared channel. No guarantees or predictions can be given here on when a node is allowed to send. In contrast, QoS routing is a mechanism under which the feasible paths provide QoS guarantees. The goal of QoS routing is to identify paths that have sufficient resources to satisfy a set of constraints and at the same time achieve global efficiency in resource utilization. The path selection in a wireless ad hoc network must be realized in an automatic and distributed way.

QoS routing consists of two parts – routing algorithm and routing protocol [7]. The routing algorithm is used to compute appropriate feasible paths while the routing protocol consists of all actions that inform the individual nodes with a consistent and updated view of the network. The routing protocol must typically define the information that is helpful in taking routing decisions and the way the information is communicated to the nodes and encoded in the routing table. The behavior of the routing protocol drives the network dynamics and critically affects the performance of the routing algorithm. The routing table is the local database of routing information and defines for each destination and for each path, the cost associated on the selection of the specific node as the next hop to forward data to the desired destination. The routing algorithm makes use of this information to actually select the paths and forward data along them.

The routing protocol can follow two different design approaches – bottom-up and top-down [8]. Many classical routing protocols follow a top-down approach. In a typical top-down design, a centralized algorithm is implemented in a distributed system. The centralized algorithm calculates the shortest path to the destination based on the knowledge of the global state of the network. Thus, the modification of the centralized algorithm has to take into consideration the intrinsic limitations of the distributed architecture in terms of limited state knowledge and delays in the propagation of the information. As a result of these modifications, several properties of the centralized algorithm may be rendered unsuitable for the wireless network. This approach requires asserting some general formal properties of the system.

In bottom-up approach, there are two main steps (1) A well defined protocol specification of individual nodes for interaction with one another and with the environment, and (2) The evolution of global behavior of the network as a result of these local level interactions. The bottom-up design approach is generally more flexible, scalable and capable of adapting to variety of situations. The negative aspect is that it is hard to state the formal properties and the expected behavior of the system.

Routing Protocol: The routing protocol presents correct and updated information of the topology and states of the link. The information about the state can be at three levels – local, global and aggregated [9]. In the local state representation each node maintains an up-to-date information of the node state and the link state. Distance vector protocols make use of the local state representation. In global state representation each node maintains a complete topological database of the network. The link state protocols exploit this strategy. The totality of the local state information for all nodes constitutes the global state information and is constructed by exchanging the local sate information for every node among all the network nodes at appropriate intervals. The global state is just an approximation of the actual state. The distribution of the state information can highly increase the communication overhead. The global state algorithms converge faster but require more CPU power and memory than the local state algorithms. To reduce the protocol overhead associated with the frequent distribution of update, aggregated global state is proposed. It is obtained by first partitioning the network into hierarchical clusters and then aggregating the information. Such aggregation represents partially true global state [4] but scales better.

Routing Algorithm: The routing algorithm is static and is used to make a routing decision to find a feasible path based on the collected information. The classification of the routing algorithms can be based on many criteria.

Based on behavior:
Reactive protocols gather routing information only in response to an event which may be due to the requirement of a new route or due to the failure of an existing route. Examples are ad hoc on demand distance vector (AODV) [5] routing protocol, and dynamic source routing (DSR) [6]. Using this approach, better route can be computed as the knowledge of most recent link characteristics and the exact QoS requirements is available. Path route caching is also not required. The negative aspect is that overhead is high due to per-request processing.

In proactive protocols, the information is constantly gathered so that it is readily available when it is required. Examples are destination sequenced distance vector (DSDV) [10] routing and wireless routing protocol (WRP). The advantage of this approach is that the proactive gathering of routing information can be used to build statistical estimates of the relevant aspects of the network dynamics that can be used to learn and adapt with continuity the local routing policies. On the other hand, it is infeasible to build sound estimates when using a purely reactive strategy since there is no continuity of information gathering.

Both of above approaches focus on finding the shortest path between the source and the destination by considering the node status and network configuration when a route is desired. Constraint based routing approach use metrics other than the shortest path to find a suitable and feasible route. Examples are Associativity-based routing (ABR) [11] which takes into account the node's signal strength to choose the path.

Based on Selection Rule:
The deterministic algorithms use a deterministic selection rule applied to the information contained in the routing table to decide the next hop. Usually it relies on the greedy selection of best routing alternative. The probabilistic algorithms make use of the probabilistic selection rule. It results in suboptimal choices but spreads the traffic across different concurrent paths resulting in load balancing. The probabilistic scheme requires more computational power and memory resources to process each packet and maintain the necessary routing information. The advantage is that due to a certain level of randomness in the selection rule, it adds robustness and flexibility to the routing system to better cope up with the network variability.

Based on routing decision:
In source routing, a feasible path is locally computed at the source node using the global information stored at the node. The path information is then disseminated to other nodes. In this, loop free path is guaranteed. However, global state information is required to be maintained at each node. In distributed or hop-by-hop routing, the source as well as other nodes is involved in path computation by identifying the adjacent nodes to which the packet must be forwarded. In this, the inconsistency in the routing tables may create loops. Hierarchical routing uses the partial global information to determine a feasible path using source routing.

2.1. Challenges in QoS Routing

QoS routing needs to take into account both the applications requirements and the availability of network resources. As a result, QoS routing in ad hoc networks impose great challenges.
1. Dynamic topology: The topology of wireless ad hoc networks may vary with time. This leads to imprecise network state information at the nodes and thus makes it difficult to provide QoS guarantees. In fact, when the network topology changes at a fast rate, it would not be possible to provide any QoS guarantees.
2. Bandwidth sharing: The bandwidth in a wireless network is limited and is also shared by other nodes in the network. The transmission from a node not only consumes local resources but also consumes the bandwidth of the neighbors within the contention range. Thus resource allocation for QoS is difficult.
3. Resource constraint: Mobile nodes have limited power supply. This limited power supply should be used in a manner that prolongs the lifetime of the battery. If the battery power is used blindly, mobile nodes will fail quickly which will affect the network availability and functionality.
4. Lack of centralized control: The absence of any centralized control demands the routing protocols to be self-creating and self-organizing. Further, the protocols must also be distributed in nature.

The QoS flows require certain resources to be reserved. The imprecise network state information and lack of centralized control makes it difficult to provide Hard QoS. In other words, QoS requirements are not guaranteed to be met for the entire session. The focus of the entire research community is to provide Soft QoS [13] or better than best-effort service in wireless ad hoc networks. Soft QoS, implies that the failure to meet the QoS requirement is allowed. It means that after connection setup, there may exist transient periods of time when the QoS specification is not honored. However, a minimum level of QoS satisfaction must be guaranteed by the network. The QoS satisfaction is quantified as the ratio of total disruption time to the total connection time. This ratio should not be higher than a threshold [14].The actual QoS for the session may vary between a range specified by a minimum and maximum. The mobility of the nodes adds another dimension of difficulty. If the topology changes too frequently, then providing even Soft QoS is not possible. Therefore, combinatorial stability [12] must be met to provide QoS guarantees. It means that given a specific time window, topology changes occur sufficiently slowly to allow successful propagation of topology updates as necessary.

3. QoS Metrics

QoS constraints are specified in terms of end-to-end performance, such as delay, bandwidth, probability of packet loss, delay jitter, etc. Although loss probability, cost, and delay jitter are useful QoS metrics, delay and bandwidth are the two most important QoS metrics. In general, the QoS metrics could be concave, multiplicative, or additive. Let m be the performance metric for the link (u, v) connecting node u to node v, and path $(u, u_1, u_2, \dots, u_k, v)$ a sequence of links for the path P from u to v. Three types of constraints on the path can be identified [15]:

1. *Additive constraint*: A constraint is additive if
$$m(P) = m(u, u_1) + m(u_1, u_2) + \dots + m(u_k, v)$$
For example, the end-to-end delay is an additive constraint because it consists of the summation of delays for each link along the path.

2. *Multiplicative constraint*: A constraint is multiplicative if
$$m(P) = m(u, u_1) m(u_1, u_2) \dots m(u_k, v)$$
The probability of a packet sent from a node u to reach a node v, is multiplicative, because it is the product of individual probabilities along the path.

3. *Concave constraint*: A constraint is concave if
$$m(P) = \min\{m(u, u_1), m(u_1, u_2), \dots m(u_k, v)\}$$

The bandwidth requirement for a path between node u and v is concave because it consists of the minimum bandwidth between the links along the path. The basic QoS routing problem can be classified into four classes depending on the metric [9]. If the path metric is optimized, then the problem is called link optimization for concave metric and path optimization for additive/multiplicative metric. Similarly if the path metric is constrained, then the problem is called link constrained for concave and path constrained for additive / multiplicative metric. Possible composite routing problems can be obtained from these four basic QoS routing. Examples are link constrained link optimization, link constrained path optimization, path constrained path optimization, etc. The composite QoS routing problems involving two or more additive/multiplicative metrics is NP complete whereas the rest is solvable in polynomial time [15]. To find the path with multiple constraints, the commonly used approach is sequential filtering where the paths based on the primary constraint are selected. The primary set is optimized by eliminating the paths not satisfying the secondary constraint. The formal definition of Multi-Constrained Path (MCP) problem is given later. The MCP problem can have two or more conflicting objectives which leads to *nondominated* solutions [16].

3.1. Multi Constrained Path (MCP) Problem

Consider a network topology modeled by a graph $G = (V, E)$ where V is the set of nodes is and E is the set of edges. Each link $(u, v) \in E$ is characterized by m additive QoS metrics. An m-dimensional non-negative weight vector is associated with the link $w_i(u, v)$ for $i = 1, 2, \dots, m$. Let the m positive path constraints be given by L. The MCP problem is given as

$$w_i(P) = \sum_{(u,v) \in P} w_i(u, v) \leq L_i \qquad \qquad \text{(1)}$$

where $i = 1, 2, \dots, mm$. The paths that satisfy the above condition are feasible paths.

3.2. Multi Constrained Optimal Path (MCOP) Problem

There may be many paths that satisfy the constraint specified by Eq. (1). The problem that additionally optimizes some length function *l(P)* is called the Multi-Constrained

Optimal Path. Formally, the problem is to find a path P from a source node s to a destination node d such that

$$w_i(P^*) \leq L_i \ \ for \ i = 1,2,...,m$$

where $l(P^*) \leq l(P) \ \ \forall P$

3.3. Path Dominance

Consider the number of constraints as 2 (i.e. m=2). Let there be two paths P_1 and P_2 from the source to the destination. Each path is characterized by a path weight vector $(w_1(P_1), w_2(P_1))$ and $(w_1(P_2), w_2(P_2))$. If P_1 is shorter than P_2, then $w_i(P_1) < w_i(P_2)$ for all $1 \leq i \leq m$. In that case, any path from the source to the destination that uses P_1 will always be shorter than P_2. In another scenario, $w_i(P_1) \leq w_i(P_2)$ for some indices i and $w_j(P_1) > w_j(P_2)$ for atleast one index j. In this case, the two points, are said to be nondominated paths. In summary, path P is called nondominated, if there does not exist a path P' for which $w_i(P') \leq w_i(P)$ for all link weight components i, except for at least one j for which $w_j(P') < w_j(P)$. The set of nondominated solution constitute the pareto optimal set as shown in fig 1.

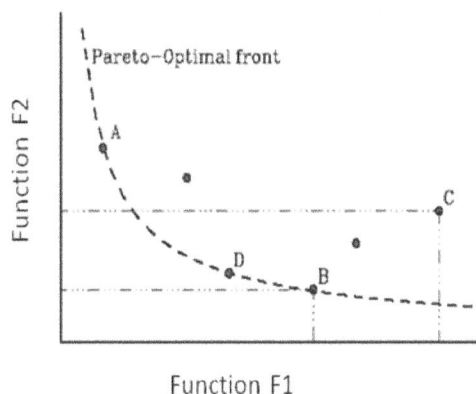

Fig. 1 Pareto set for MCP and MCOP problem

4. QoS AWARE ROUTING PROTOCOLS

Any routing protocol can be verified for its correctness and effectiveness. This requires a formal characterization and quantification of the routing protocol. We provide the analysis of some QoS aware routing protocols (illustrated in Table 1) based on the following factors.

(a) Type of algorithm
    a. Unicast – This has exactly one sender and one receiver
    b. Multicast – In this, the number of senders may be one or more. The senders address the packets to multiple destinations.
    c. Broadcast – In this, the packets are addressed to all the nodes in the network.
(b) Network Architecture:
    a. Flat – In this all nodes are at equal level and maintain the routing table. Thus every node is equally responsible for routing.
    b. Hierarchical – The nodes designated as clusterheads form a virtual backbone and maintain the routing table.
(c) Quality of Service
    a. Constraints – This specifies the number of constraints used for route selection. The routing algorithm may be single or multiconstrained.

    b.   Metrics – The QoS requirements are specified in terms of metrics. The metrics may be additive, multiplicative or concave. The most commonly used metric are bandwidth and delay.

(d) Routing protocol complexity

    a.   Communication complexity – This relates to the exchange of information to have an up-to-date information of the network topology and assist in finding a feasible path satisfying the constraints.

    b.   Memory Complexity – This denotes the amount of memory that is required to be stored to determine the state information.

(e) Routing Algorithm

    a.   Routing Type :Source Routing, Distributed Routing, Hierarchical Routing

    b.   Route Discovery: Proactive or Table Driven, Reactive or On Demand, Hybrid Routing

    c.   Routing Complexity – This indicates the number of messages needed to discover a feasible route satisfying the constraints.

    d.   Routing Overhead – This specifies the control packets to be exchanged for finding a feasible path.

    e.   Route Caching – This parameter indicates whether the routing algorithm allows route caching or not. If route caching is not allowed then all the communication will be direct from source to destination and the intermediate nodes are not allowed to send the route reply packets.

(f) Resource Estimation – This estimates the availability of resources in the network. The information is used to find a feasible path satisfying the constraints.

(g) Resource Reservation – This specifies whether the routing algorithm just determines a feasible path or takes into account the reservation of the path. The reservation may be hard or soft.

(h) Route Maintenance – The mobility of nodes may cause frequent route breakages. The methods used for route maintenance are

    a.   Route Prediction – This specifies whether the algorithm has a prediction scheme to predict the breakage of route.

    b.   Redundant Routes – The algorithm may find more than one route for route maintenance. One route acts as the primary route and the other routes act as secondary routes.

4.1. Core Extraction Distributed Ad Hoc Routing (CEDAR) [17]

CEDAR is a routing protocols designed for small to medium sized networks. It dynamically establishes a core set for route set up, QoS provisioning, routing data and route maintenance. The link states of stable high bandwidth links are propagated to the core nodes. Route computation is on demand and is performed using only local state. Two assumptions are made in CEDAR: (1) The MAC/link layer can estimate the available link bandwidth and (2) The network consist of tens to hundreds of nodes. The key components in CEDAR are:

Core extraction – A set of nodes is elected to form the core that maintains the local topology, topology and available information exchange, perform route discovery and route maintenance on behalf of all the nodes in its domain. A greedy algorithm is used to create an approximate dominating set proactively. All hosts in the set are either member of the core or one-hop neighbors of core hosts.

State propagation - The bandwidth availability information of stable links is disseminated to other nodes using increase and decrease waves mechanism. The waves are generated when an estimate of the available bandwidth of a core node is changed by a certain amount. The information about the small changes in available bandwidth is kept locally whereas the relatively stable bandwidth information is propagated among the core hosts. The increase in the available bandwidth of the core node is represented by increase waves. This information is propagated periodically. The decrease wave which represents the decrease in available bandwidth is propagated immediately to prevent the overestimation of the available bandwidth by the core nodes.

Route computation - The route computation first establishes a core path from the domain of the source to the domain of the destination. For route selection, the shortest widest route is chosen among all the feasible routes using a two-phase Dijikstra algorithm. The directional information provided by the core path is used iteratively to find a partial route from the source to the domain of the furthest possible node in the core path satisfying the bandwidth constraints. This node then becomes the source for the next iteration.

Route maintenance – The algorithm uses two strategies for route maintenance – rerouting and repairing. In rerouting, the source is notified of the failure and a new route is computed to reach the destination. In repairing, the link is locally repaired at the point of failure by the surrounding nodes.

The routing overhead in CEDAR is very low due to the presence of core nodes. The increase and decrease wave mechanism for link state propagation also ensures that the available bandwidth information is propagated without incurring a high overhead.

4.2. Ticket Based Probing (TBP) [13]

TBP is a distributed ticket based routing proposed by Chen and Nahrstedt. The basic idea of the TBP is to utilize tickets to limit the number of paths searched during route discovery. A ticket is the permission to search a single path. The tickets are used to find delay constrained or bandwidth constrained routes. In this, when a feasible route is to be established between the source and the destination then the source sends a limited number of probes (routing messages) to some neighboring nodes. Each probe contains at least one ticket. When the connection requirements are tighter, the probes may carry more than one ticket. The total number of tickets for path discovery is constant. At an intermediate node, a probe with more than one ticket may be split into multiple probes with tickets distributed between them. Each probe searches for a different downstream subpath. The decision to split the probe and to which neighbors the probes should be forwarded is taken on the basis of its available state information. TBP is based on the assumption that stable links tend to remain stable in contrast to the transient links. Each node i collects statistical information about delay $D(t)$, bandwidth $B(t)$ and cost $C(t)$ for all other nodes in the network using distance vector. Along with the delay and bandwidth each node also maintains the associated variation $\Delta D(t)$ and $\Delta B(t)$ by which the next reported value will differ with the current one. The protocol searches for a least cost delay constrained path or least cost bandwidth constrained path. For this purpose, it defines two types of tickets – yellow and green. The number of these tickets is based on the imprecise information and look for feasible paths or least constrained path. In the case of route failures, TBP utilizes three mechanisms — path rerouting, path redundancy, and path repairing. Rerouting requires that the source node be informed of a path failure. In path rerouting, the source node is informed about the route failure which then initiates a route discovery to find a new path to the destination. In path redundancy, multiple routes are established from the source to the destination. One path may serve as the primary path and the other paths may be used as backup paths. The resources are reserved only along the primary path to reduce the wastage of resources. In path repairing mechanism, TBP tries to repair the route at the point of failure.

4.3. Ad Hoc QoS On-Demand Routing (AQOR) [18]

Ad hoc QoS On-demand Routing (AQOR) is an on demand QoS aware routing protocol. It integrates (1) bandwidth estimation and end-to-end delay measurement in the route discovery process, (2) bandwidth reservation, and (3) adaptive route recovery. The protocol uses limited flooding to discover the best route available in terms of smallest end-to-end delay and bandwidth guarantee.

The bandwidth estimation is accomplished by disseminating the traffic information to neighbors through periodic announcement packets, called Hello packets. Each node i includes the self traffic in the hello packets. Typically, it consists of three types of traffic (1) Self-traffic $(B_i(self))$ i.e. the traffic between the node i and its neighbors. This traffic indicates the bandwidth consumed by the traffic transmitted or received by node i, (2) Neighborhood traffic $(B_i(neighborhood))$ i.e. the total traffic between i's neighbors and (3) Boundary traffic $(B_i(boundary))$ i.e. the total traffic between i's neighbors and nodes that are outside i's range. The sum of the neighbors' traffic of a node is estimated as the total traffic affecting the node

$$B_i(agg) = B_i(self) + B_i(neighborhood) + B_i(boundary)$$

However, the estimated traffic can be larger than the real overall traffic. The available bandwidth is thus a lower bound on the real available bandwidth and imposes stringent bandwidth admission control threshold.

The route discovery process is activated when a route is desired. The route request packet includes both bandwidth and end-to-end delay constraints. Upon receiving the route request packet, the intermediate nodes perform bandwidth admission hop-by-hop. If the intermediate nodes have sufficient bandwidth, the request is accepted and the node adds the route in its routing table with an expiration time. The status of the node is set as explored and the request is rebroadcast to the next hop. The node remains in the explored status for a period of 2D where D is the end-to-end delay. If the reply packet does not arrive within the expiration time, the entry is deleted. The reply packet that arrive late are ignored to reduce overhead and exclude invalid information from the routing table. The smallest delay route with sufficient bandwidth is chosen as the route satisfying the QoS constraints.

The bandwidth reservation is made along the route discovered, but it is activated only when the data flow passes through the route. The reservation is soft and if the node does not receive data packets for a certain time interval, the node immediately invalidates the reservation. The adaptive route recovery consists of QoS violation detection. The end-to-end QoS violations are caused either by congestion or route breakage. These violations are detected by the destination which then initiates a destination initiated route recovery. The end-to-end delay violation is also detected by the destination by monitoring the delay of the arriving packets. If the delay exceeds the maximum delay requirement, QoS recovery is triggered. The route failure or network partition is detected by the neighbor lost detection mechanism. The non arrival of the Hello message in time indicates a route failure. When a neighbor lost is detected, the source is notified about the break which then initiates the reroute process. The QoS violations are detected by the destination.

4.4. Bandwidth Estimation QoS Routing (BEQR) [21]

The aim of this protocol is to provide Soft QoS or better-than-best effort service rather than guaranteed service. The design of the protocol is based on two schemes (1) feedback scheme i.e. providing feedback about the available bandwidth to the application, and (2) admission scheme i.e. admit a flow with the requested bandwidth. Both of these require the estimation of the bandwidth. Thus, bandwidth estimation is the key in the design of BEQR. The route discovery function of this protocol is based on AODV [5] with a modified packet format. The residual bandwidth can be estimated either by listening to the channel and calculating the ratio of free and busy times or by appending the node's current bandwidth with that of its 1-hop neighbors to AODV's periodic hello messages. AODV's route request packets (RREQ) include additional information about the used scheme and either the bandwidth constraint or the minimum of bandwidth constraint and detected bandwidth on the partial path.

4.5. Adaptive Proportional Routing (APR) [23]

The authors in [23] argue that majority of the QoS routing schemes obtain a global view of the network state. This gives rise to prohibitive communication and processing overheads. APR proposes a localized approach to QoS routing. In this, the source nodes infer the network QoS state on flow blocking statistics collected locally and perform flow routing using this localized view of the network QoS state. For each source-destination pair, the protocol sets up one or multiple explicit routed paths *a priori* using MPLS (Multi-Protocol Label Switching). These paths are referred as candidate paths. Each flow routed along the candidate path has a certain probability of being blocked. The virtual capacity of this path is computed using the knowledge of capacity and blocking rates. APR tries to equally distribute the flows among the available paths w.r.t the virtual capacity of each path selecting the shortest paths. No QoS information is exchanged between the nodes; thus reducing the protocol overhead.

4.6. QoS Enabled AODV (QAODV) [24]

QoS enabled AODV was proposed by Perkins *et al.* [24]. To support QoS, the protocol extends the formats of the RouteRequest and RouteReply packets. This necessitates the need to modify the routing table structure. Four new fields are appended in the routing table for QoS support – (1) Maximum delay (2) Minimum available bandwidth (3) List of sources requesting delay guarantees and (4) List of sources requesting bandwidth guarantees. The QoS constraint is specified in the RouteRequest packets. The intermediate nodes forward the RouteRequest packets if the requested QoS parameter can be satisfied, otherwise the packet is dropped. The actual parameter value that can be satisfied for the path is recorded in the routing table. The intermediate nodes may generate the QoS-LOST message packet if the node detects that the required QoS constraint cannot be satisfied any more. This message is transmitted to all the nodes along the path.

5. CROSS LAYER DESIGN

The traditional network architectures are designed using layering approach. In this approach, the entire network communication functionality is divided into modules known as layers. Each layer fulfils a limited and well defined purpose. Every layer of the system is designed separately and is independent of the application. Every layer offers services to the layer above it and also accepts the services of the layer below it. The information exchange between the layers is only through interfaces which are limited. Thus, the information exchange and coupling between the layers is kept as low as possible. The advantage of this approach is modularity and simplicity. Also the same network architecture can be used by many applications. The negative aspect is that the protocols designed using the layered approach is not optimal for any application. The wireless network characteristics are quite different from the wired networks; so are the challenges. The wireless channel characteristics generally affect all the OSI layers. Optimizing each layer individually to fix the problem leads to unsatisfactory results. It is argued in [19] that it is hard to achieve design goals like energy efficiency and QoS using the traditional layered approach. In other words, a cross layer design (CLD) is needed to achieve the optimal results. Cross layer design seeks to enhance the performance of the system by jointly optimizing multiple protocol layers [22]. The extreme design alternative for CLD is to have a complete layerless approach i.e. collapsing the entire stack to obtain completely integrated protocol architecture. In the other approach some layers of the protocol stack can be merged to obtain the desired results. Since the modular approach has proven itself over time, in yet another approach for CLD, the layers are kept intact and the information is shared between the layers either directly or through a database. Cross layer design is becoming an integral part of several wireless standards. For example 3G standards such as CDMA 2000, Broadband Radio Access Network (BRAN) of HiperLAN2, High Speed Downlink Packet Access (HS-DPA) of 3G Partnership Project and IEEE Study Group Mobile Broadband Wireless Access Networks [2] are based on cross layer design. In designing an architecture using cross layer approach, one has to keep in mind that CLDs without solid architecture guidelines can inevitably lead to "Spaghetti Design"[20]. CLD with tight coupling between the layers becomes hard to review and redesign. The change in one subsystem implies changes in other parts as everything is connected. This may lead to unpredictable systems as it is hard to forsee the impact of modifications.

5. CONCLUSION

Wireless ad hoc networks are likely to be the center of future communication. Although it is very difficult but providing QoS guarantees has become essential for the operation of today's multimedia wireless networks. This paper presented an overview of the QoS routing protocols and outlined the challenges that make QoS routing difficult in wireless ad hoc networks. We also presented an extensive review of some current existing protocols. However, several important issues remain to addressed before QoS in wireless ad hoc networks becomes a reality. The energy constraint is of principle interest in wireless ad hoc networks. The future direction for research is to take into consideration the battery constraint while providing QoS. Designing such QoS protocols that optimize multiple objectives is computationally intractable. Thus, a multiobjective protocol architecture design for providing QoS and minimizing energy dissipation has to be thoroughly investigated. The current paradigm shift is towards cross layer optimization to provide energy efficient QoS solutions. Also providing QoS in broadcasting and multicasting has found little attention in the literature.

REFERENCES

[1]   Wu, J. & Stojmenovic, I., (2004) "Ad Hoc Networks", IEEE *Computer*, Vol. 37, No. 2, pp. 29-31.

[2]   B. Tavli and W. Heinzelman, (2006) "Mobile Ad Hoc Networks" Springer Publishers.

[3]   Crawley, E., "A Framework of QoS Based Routing in the Internet", RFC 2386, http://www.ietf.org/rfc/rfc2386.txt

[4]   Chen, S., (1999) "Routing Support for Providing Guaranteed End-to-End Quality of Service", Ph. D Thesis, University of Illinois, Urbana-Champaign.

[5]   Perkins, C. E., Royer, E. M. & Das, S. R., (2002) "Ad Hoc On-Demand Distance Vector (AODV) Routing", draft-ietf-manet-aodv-10.txt, IETF MANET working group.

[6]   Johnson, D.B., Maltz, D.A. & Hu, Y.C., (2002) "The Dynamic Source Routing Protocol for Mobile Ad Hoc Networks", draft-ietf-manet-dsr-07.txt, IETF MANET working group.

[7]   Masip-Bruin, X, & Kuipers, F., *et al.*, (2006) "Research Challenges in QoS Routing", Elsevier Journal of Computer Communication, Vol. 29, No. 5, pp. 563-581.

[8]   Farooq, M., & Di Caro, G.A., (2008) "Routing Protocols for Next Generation Networks Inspired by Collective Behaviors of Insect Societies: An Overview", Swarm Intelligence (Natural Computing Series), Springer, pp. 101-160.

[9]   Chen, S., & Nahrstedt, K., (1998) "An Overview of Quality of Service Routing for Next Generation High Speed Networks: Problems and Solutions", *IEEE Network,* Vol. 12, No. 6, pp. 64-79.

[10]   Perkins, C.E., & Bhagwat, P., (1994) "Highly dynamic destination-sequenced distance vector routing (DSDV) for mobile computers", Computer Communication Review, Vol. 24, No. 4, pp. 234-244.

[11]   Toh, C.K., (1997) "Associativity Based Routing for Ad Hoc Networks" Wireless Personal Communications, Vol. 4, No. 2, pp. 1-36.

[12]   Chakrabarti, S., (2003) "Quality of Service in Mobile Ad Hoc Networks", Handbook of Ad Hoc Wireless Networks, CRC press.

[13]   Chen S., & Nahrstedt, K., (1999) "Distributed Quality of Service in Ad Hoc Networks", IEEE Journal on Selected Areas in Communication", Vol. 17, No. 8, 1999, pp 1-18.

[14]   Mahapatra, P., Li, J., & Gui, C., (2003) "QoS in Mobile Ad Hoc Networks", IEEE Wireless Communications, Vol. 10, No. 3, pp. 44-52.

[15]   Wang Z., & Crowcraft, J., (2002) "Quality of Service Routing for Supporting Multimedia Applications", IEEE Journal on Selected Areas in Communications, Vol. 14 , No. 7, pp. 244-256.

[16]   Meigham P., & Kuipers, F., (2004) "Concept of Exact QoS Routing Algorithms", IEEE/ACM Transactions on Networking, Vol. 12, No. 5, pp. 851-864.

[17]   Sivakumar, R., Sinha, P., & Bharghavan, V., (1999) "CEDAR: A Core-Extraction Distributed Ad Hoc Routing Algorithm", IEEE Journal on Selected Areas in Communications, Vol. 17, No. 8, pp. 1454-1465.

[18]   Xue Q., & Ganz, A., (2003) "Ad Hoc QoS On Demand Routing (AQOR) in Mobile Ad Hoc Networks", Elsevier Journal of Parallel and Distributed Computing, Vol. 63, No. 3, pp. 154-165.

[19]   Heinzelman, W., (2000) "Application Specific Protocol Architectures for Wireless Networks", Ph.D Thesis, Massachusetts Institute of Technology, 2000

[20]   Kawadia V., and Kumar, P.R., (2005) "A Cautionary Perspective on Cross Layer Design", IEEE Wireless Communication, Vol. 12, No. 1, pp. 3-11.

[21]   Chen L., & Heinzelman, W., (2005) "QoS Aware Routing Based on Bandwidth Estimation for Mobile Ad Hoc Networks", IEEE Journal on Selected Areas in Communication, Vol. 23, No. 3, pp. 561-572.

[22]   Van der Schaar M., & Sai Shankar, N., (2005) "Cross layer wireless multimedia transmission – challenges, principles and new paradigms", IEEE Wireless Communication, Vol. 12, No. 4, pp. 50-58.

[23]   Nelakuditi, S., Zhi-Li Zhang, Tsang, Rose P., & David H. C. Du, (2002) "Adaptive Proportional Routing: A Localized QoS Routing Approach", IEEE/ACM Transactions of Networking, Vol 10, No. 6, pp. 790-804.

[24]   Perkins, C., & Royer, E., (2003) "Quality of Service for Ad Hoc on-demand Distance Vector Routing", Internet Draft, draft-perkins-manet-aodvqos-02-txt.

Table 1 : Comparison of Routing Protocols

Parameters	TBP [13]	AQOR [18]	CEDAR [17]	BEQR [21]	APR [23]	QAODV [24]
Type of Algorithm	Unicast	Unicast	Unicast	Unicast	Unicast	Unicast
Network Architecture	Flat	Flat	Hierarchical	Flat	Flat	Flat
Quality of Service						
a. Constraints	Single constrained	Multi-constrained	Single Constrained	Single Constrained	Single Constrained	Single Constrained
b. Metrics	BW or Delay	BW and Delay	BW	BW	BW	BW or delay
Routing Protocol Complexity						
a. Communication Complexity	$O(n)$	$O(n)$ per second	$O(n)$	$O(n)$	Not Available	$O(n)$
b. Space Complexity	$O(n^2)$	$O(n)$	$O(n^2)$	$O(n)$	$O(p)$	$O(n)$
Routing Algorithm						
a. Routing Type	Distributive	Distributive	Distributive	Distributive	Source	Distributive
b. Route Discovery	Reactive	Reactive	Reactive	Reactive	Reactive	Reactive
c. Routing Complexity	$O(t \times n)$	$O(n)$	$O(n)$	$O(2n)$	Not Available	$O(2n)$
d. Routing Overhead	Limited flooding of RREQs	Full flooding of RREQs	Limited: Link state information distributed only among core nodes	Full Flooding of RREQs	Not Available	Full flooding of RREQs
e. Route Caching	No	No	Yes	No	Yes	No
Resource Estimation	No	Yes	No	Yes	No	No
Resource Reservation	Yes (soft)	Yes (soft)	Yes (soft)	No	No	No
Route Maintenance						
a. Route Prediction	No	No	No	No	No	No
b. Redundant Routes	Yes	No	No	No	No	No

Where t = no. of tickets

n = no. of nodes

p = no. of candidate paths connected to the node

5

A Simulation based Performance Evaluation of AODV, R-AODV and PHR-AODV Routing Protocols for Mobile Ad hoc Networks

Pravanjan Das[1], Sumant Kumar Mohapatra[2] and Biswa Ranjan Swain[3]

[1]Ericsson India Global Services Pvt. Ltd., Salt Lake, Kolkata, India

[2]Trident Academy of Technology, Bhubaneswar, Odisha

[3]Trident Academy of Technology, Bhubaneswar, Odisha

ABSTRACT

Mobile Ad hoc Networks (MANETs) are characterized by open structure, lack of standard infrastructure and un-accessibility to the trusted servers. The performance of various MANET routing protocols is significantly affected due to frequently changing network topology, confined network resources and security of data packets. In this paper, a simulation based performance comparison of one of the most commonly used on-demand application oriented routing protocols, AODV (Ad hoc on-demand Distance Vector) and its optimized versions R-AODV (Reverse AODV) and PHR-AODV (Path hopping based Reverse AODV) has been presented. Basically the paper evaluates these protocols based on a wide set of performance metrics by varying both the number of nodes and the nodes maximum speed. A NS-2 based simulation study shows that, as compared to AODV and PHR-AODV, R-AODV enhances the packet delivery fraction by 15-20% and reduces the latency approximately by 50%. R-AODV requires lesser node energy for data transmission.

KEYWORDS

MANET, AODV, R-AODV, PHR-AODV, packet delivery fraction, latency.

1. INTRODUCTION

A group of mobile nodes made a MANET[1]. They form a network for information exchange. For this information exchange, they never use the central authority as well as existing fixed network infrastructure. This upcoming technology creates new research opportunities and dynamic challenges for different topology of the network, bandwidth limitation, node's battery capacity improvement and multi-hop communication.

The routes are updated at regular intervals with respect to their requirement in proactive routing but routes are determined only when there is a need to transmit a data packet in reactive routing. Single path routing protocols [1-2] and multipath routing protocols [3-5] are classified on the basis of number of routes computed between source and destination.

The area of discussion in this paper is based on AODV [2], a single path routing protocol and its multipath versions, R-AODV and PHR-AODV routing protocols. AODV is both on-demand and destination initiated, that means routes are established from destination only on demand [1]. But the problem in such kind of single path routing is the increased latency and packet loss due to dynamic nature of the routing environment. Unnecessary bandwidth consumption due to periodic beacons also affects the performance of AODV.

Besides, the performance of AODV is significantly affected due to the loss of single unicasted route reply (RREP) packets. The drastically altering environment prevents the RREP packet from getting delivered to the source node. As a result the source node starts rediscovery process, which in turn increases both consumed energy and communication delay. The Reverse-AODV (R-AODV) [6], broadcasts the route reply packet throughout the network instead of unicasting it. This process generates multiple discovered partial or full disjoint paths at the source node. It also ensures both successful route discovery and data packet delivery reducing path fail correction messages.

Protecting network activity from intrusion of malicious nodes and enhancing the data security are the important issues of Mobile Ad hoc networks. Sometimes the performance of R-AODV routing protocol gets significantly affected due to the activity of these active malicious nodes. So PHR-AODV [11], builds multipath to destination and adaptively hops between the available paths for transmission of data packets. As a result load distribution arises and it ensures that the nodes do not get depleted of energy which in turn increases the network lifetime.

In this paper, a significant amount of network parameters and energy related parameters have been considered in order to compare the above mentioned routing protocols. The graphs obtained are based on multiple readings and later averaging for a single plot point. The simulation environment considered in this paper is highly dynamic (lesser pause time and higher nodes speed) and the simulation software used is network simulator (NS-2).

2. OVERVIEW OF ROUTING PROTOCOLS

This section briefly describes AODV, R-AODV and PHR-AODV routing protocols.

2.1. AODV Routing Protocol

The Ad hoc On-Demand Distance Vector (AODV) [1-2] is a destination initiated routing protocol that maintains the sequence number concept for loop free routing and initiates the route discovery process on demand, hence has the combined features of DSDV and DSR respectively. The entire working principle of AODV can be covered under two important phases: Route Discovery and Route Maintenance.

The node disseminates and avoids repeated processing of RREQ packets at nodes by matching the source IP address and RREQ ID pair of the packet with nodes stored information. A RREP packet is generated by a node if it is itself the destination node or it has an active route to the destination. The destination node unicasts the RREP packet back towards the source node along the reverse path.

Another route error known as RERR message initiates and defects a link break for for the next hop of an active route in it's routing table but it is not attempting for any other local repairing. As a decision the RERR message which is received looks for an another route from it's routing table.

2.2. R-AODV Routing Protocol

The Reverse AODV (R-AODV) [6], an optimized AODV routing protocol uses a reverse route discovery methodology in order to avoid RREP packet loss. R-AODV prevents a large number of retransmissions of RREQ packets which in turn reduces the congestion in the network and enhances the delivery ratio [7].

From the RREQ packet transmission point of view, both AODV and R-AODV play the same role. The RREQ packet format of R-AODV is same as that of AODV. When the destination node receives the first RREQ packet, a reverse route request (R-RREQ) packet is initiated and broadcasted to its neighbour nodes within its transmission range. The R-RREQ packet format is given in Table 1.

Table1. R-RREQ Message Format [6]

Type	Reserved	Hop Count
Broadcast ID		
Destination IP address		
Destination Sequence Number		
Source IP address		
Reply Time		

The source node starts packet transmission after receiving the first R-RREQ message whereas the late arrived R-RREQ packets are saved for future use. A forward route entry is created if the node received R-RREQ packet is not a source node. Then the R-RREQ is broadcasted to it's neighbour nodes. After receiving the packet , it adds in a new path with a different hop or with the same next hop. This is totally depends upon a sequence number which is greater or less [7][9].

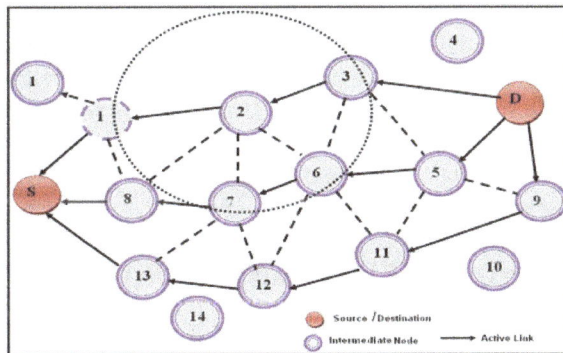

Figure1. R-RREQ from Destination to Source Node

Figure 1 shows the flooding of R-RREQ packet by the destination node in order to find the source node. So along with the primary path D-3-2-1-S, a number of paths might be built at the source node such as D-5-6-7-8-S, D-9-11-12-13-S.

2.3. PHR-AODV Routing Protocol

PHR-AODV [11] contains no permanent routes in nodes routing table. It is an extension of R-AODV routing protocol which prevents loss of data packets by active malicious node and distributes load uniformly among the nodes.

The processing of RREQ and R-RREQ packets is same as that of R-AODV. The packet formats of R-AODV and PHR-AODV are same. When the source node receives R-RREQ packets from its neighbour nodes, it simply builds partial node-disjoint paths. After receiving all the node-disjoint paths within the timeout period, source node hops between different paths (based on the ascending order of their hop count values) while sending data packets to the destination node. During this communication of data packets, if a particular path fails then that path is eliminated from the list. The source node reinitiates the route discovery when no paths remain in the list. Figure 2 shows the node disjoint paths discovered between source and destination node.

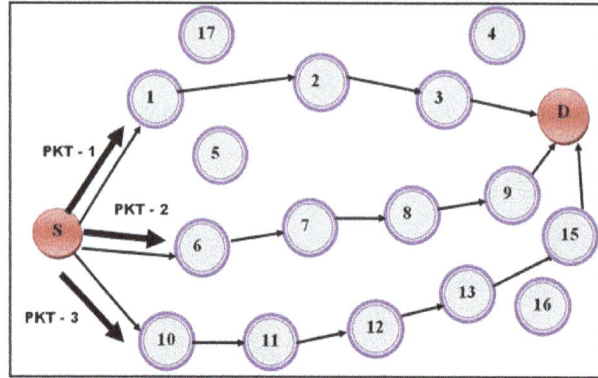

Figure2. Hopping of Paths at Source Node

PHR-AODV provides an effective and analytic method to estimate the security of the network [11]. The probability of active malicious node Pm is given as:

$$Pm = (Nrp * Nm) / Ntotal \qquad (1)$$

Where Nrp is the number of nodes in routing path, Ntotal is the number of all nodes in network, Nm is the Number of malicious nodes (2 malicious nodes are considered in this paper) and Np is the number of paths from a source to a destination. The malicious node intrusion rate is given by

$$Pi = Pm / Np \qquad (2)$$

3. SIMULATION AND PERFORMANCE EVALUATION

This section explained the simulation model and their performance metrics. This also describes the results which are analysed from the simulation.

3.1. Simulation Model

In this section, the network simulator-2 is used. It supports for simulating a multi-hop wireless ad-hoc environment completed with physical, data link and medium access control layer models. The table-2 and 3 shows the simulation parameters and energy parameters respectively.

Table 2. Energy Parameters [7][9]

Parameters	Values
Routing Protocol	AODV,R-AODV,PHR-AODV
Number of Nodes	20,30,40,50,60,70
Area of Terrain	1500 m X 1500 m
Total time for simulation	100.0 sec
Type of Traffic	CBR
Size of Packet	512 bytes
Maximum speed of Nodes	2,5,10,15,20 m/s
Bandwidth	2.0Mbps
Transmission Range	250 m
Transmission Rate	4 Packets/sec
No of flows	10
Frequency Band	2.4 GHz
Pause Time	5 sec
Propagation type	Propagation/TwoRayGround
Antenna	Antenna/OmniAntenna
Queue Type	Queue/DropTail/PriQueue
Queue Length	50
MAC type	Mac/802_11b
Mobility Model	Random Waypoint

Table 3:Simulation Parameters

3.2. Performance Metrics

Parameters	Values
Receiving Power	1.0 watt
Transmitting Power	1.4 watt
Idle Power	0.83 watt
Sleep Power	0.13 watt
Transition Power	0.2 watt
Nodes' Initial Energy	300 Joules

The following performance metrics [6-10] are considered for the simulation:

- **Packet Delivery Fraction**: It is the ratio of data packets received at the destination to those generated by the source.
- **Average End-to-End Delay**: It is the time interval between Transmitting packet and Receiving packet, which is a summation of isolation of data packets during route discovery, retransmission delays at the MAC and queuing at the interface queue.
- **Routing Overhead**: It is the sum of all control packets transmitted/received during route discovery and route maintenance.
- **Average Energy Consumed**: It is the mean value of energy consumed by a node during the whole simulation process.
- **Intrusion Rate**: It is the percentage of intrusion instances present in the test set.

3.3. Results and Discussions

The packet delivery fraction and average end-to-end delay of R-AODV characterisation are shown in figure 3 and 4. It is very obvious that above two factors improved as compared to other two protocols due to R-AODV utilises multiple recent routes at the source node which are fresh enough. Also the routing overhead of R-AODV is much larger than AODV which is clearly shown in figure 5 because R-AODV broadcasts route reply packets whereas it is unicasted in AODV.

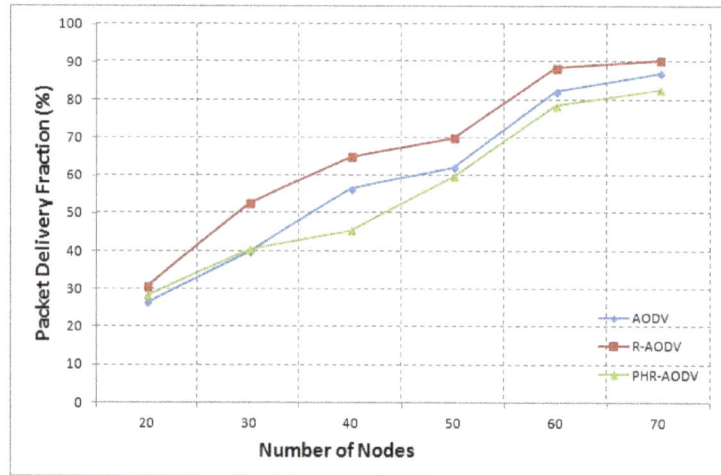

Figure 3. Packet Delivery Fraction varying Number of Nodes

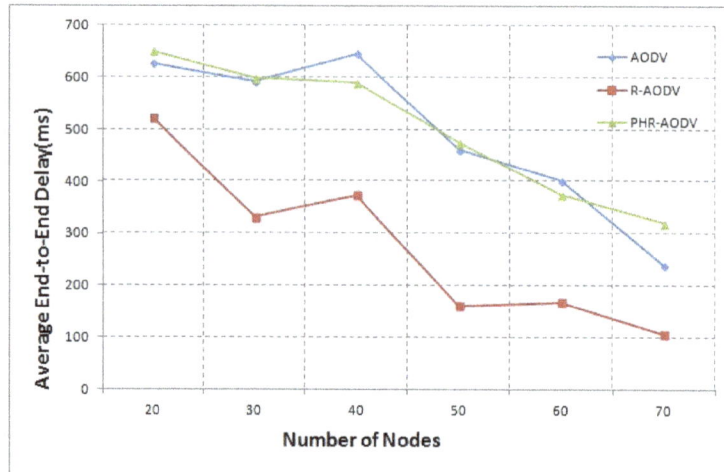

Figure 4. Average End-to-End Delay varying Number of Nodes

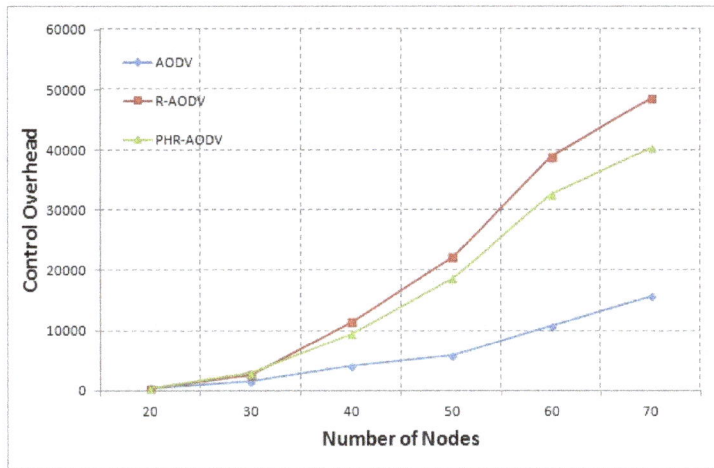

Figure 5. Control Overhead varying Number of Nodes

From Figure 6 and 7 it is clear that R-AODV completely outperforms AODV and PHR-AODV when performance is considered by varying Nodes Maximum Speed due to that R-AODV uses a number of multiple invented routes with high consistency to switch over the rapid topology changes due to higher nodes speed, which reduces the delay of communication which arises due to successful delivery of data packets.

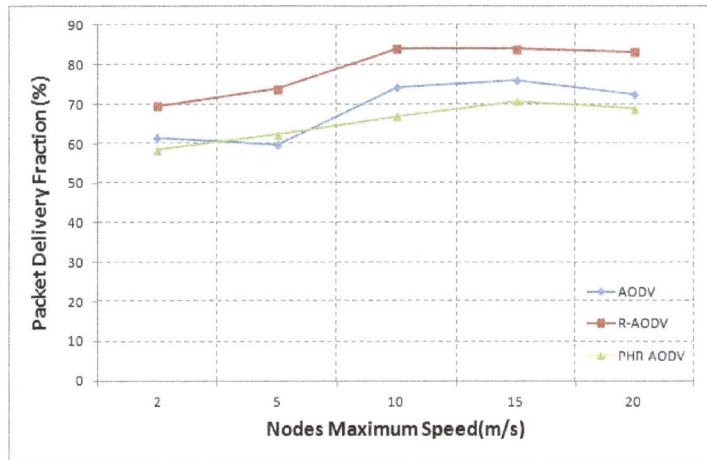

Figure 6. Packet Delivery Fraction vs. Nodes Max. Speed

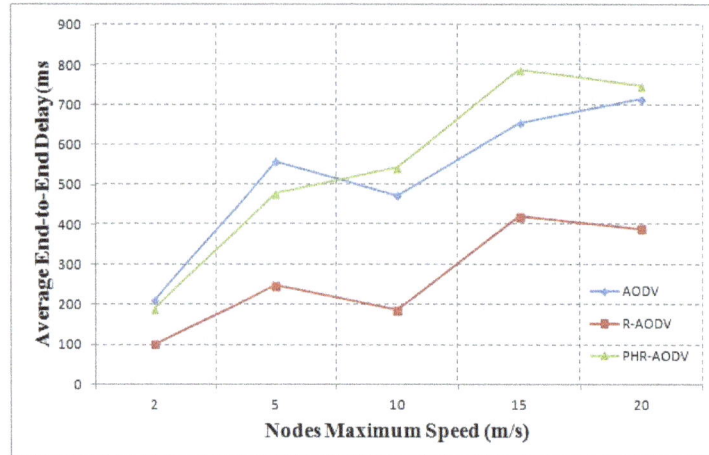

Figure 7. Average End-to-End Delay varying Nodes Max. Speed

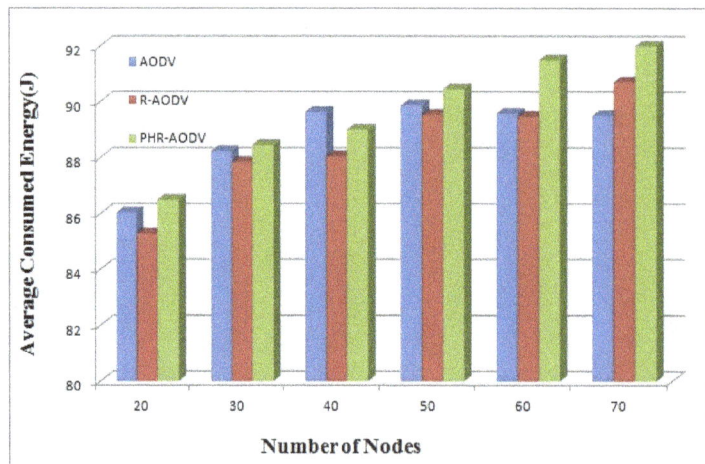

Figure 8. Average Energy Consumed varying Number of Nodes

The average consumed energy for R-AODV is lesser than AODV and PHR-AODV is clearly illustrated in figure(8). This will effective even if it has transmitted larger number of control packets than others. The reason is that the R-AODV uses fewer hops in the chosen paths to route the data packets[8]. As a result for a longer period R-AODV increases the survival of the nodes in the network.

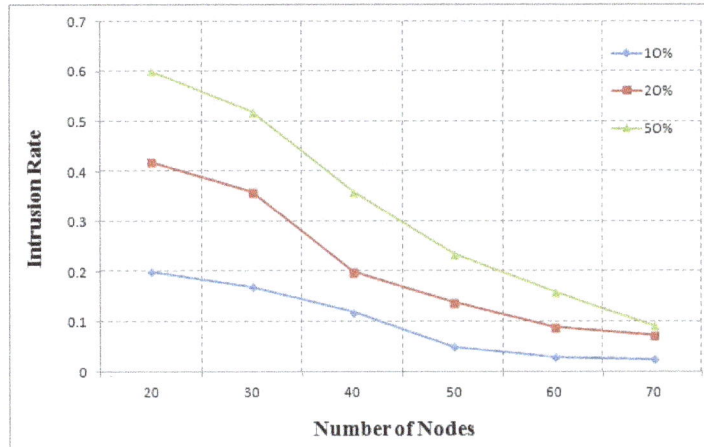

Figure 9. Intrusion Rate varying Number of Nodes

The two malicious nodes which are described in this paper, their intrusion rate decreases depending upon the number of nodes increases in case of PHR_AODV. Further the intrusion rate decreases as the probability of active malicious nodes decreases.

4. CONCLUSIONS

An extensive analytic model and a simulation study have been conducted to evaluate the performance of AODV, R-AODV and PHR-AODV using NS-2.The above simulation analysis raise a conclusion that the performance of R-AODV in terms average end-to-delay, average energy consumption and packet delivery fraction totally dominates both AODV and PHR-AODV .This is happened in high mobility scenarios at a cost of higher routing overhead. It can also be seen that in case of PHR-AODV, the intrusion rate of malicious nodes decreases with increase in number of nodes.

ACKNOWLEDGEMENTS

The authors would like to thank everyone, just everyone!

REFERENCES

[1] Elizabeth M. Royer and Chai-Keong Toh, "A Review of Current Routing Protocols for Ad Hoc Mobile Wireless Networks," IEEE Personal Communications, Vol. 6, No. 2, pp. 46-55, April 1999.

[2] C.E. Perkins and E. M. Royer, "Ad hoc on-demand distance vector routing" in Proc. WMCSA, New Orleans, LA, pp. 90–100, Feb. 1999.

[3] M. K. Marina and S. R. Das "On-Demand Multi Path Distance Vector Routing in Ad Hoc Networks" in Proc. ICNP 2001, pp. 14– 23, Nov. 2001.

[4] A. Nasipuri and S. R. Das, "On-Demand Multipath Routing for Mobile Ad Hoc Networks" Proc. ICCN 1999, pp. 64–70, Oct. 1999.

[5] Tarique, M., Tepe, E., Sasan Adibi, and Shervin Erfani, "Survey of multipath routing protocols for Mobile Ad-hoc networks", Journal of Network and Computer Applications, 32:1125-1143, 2009.

[6] Chonggun Kim, Elmurod Talipov and Byoungchul Ahn, "A Reverse AODV Routing Protocol in Ad Hoc Mobile Networks," LNCS 4097, pp. 522-531, 2006.

[7] Pravanjan Das and Upena D Dalal, "A Comparative Analysis of AODV and R-   AODV   Routing Protocols in MANETS", International Journal of Computer   Applications   72(21):1-5,   June 2013.

[8] Khafaei Taleb and Khafaie Behzad, "The Effect of Number of Hops per Path on Remind Energy in MANETs Routing Protocols," International Journal of Computer Applications vol. 43, no. 24, pp. 23-28, April 2012.

[9] Humaira Nishat, Vamsi Krishna K, D.Srinivasa Rao and Shakeel Ahmed, "Performance Evaluation of On-Demand Routing Protocols AODV and Modified AODV (R-AODV) in MANETS," International Journal of Distributed and Parallel Systems, vol. 2, no. 1, January 2011.

[10] Talipov, Elmurod, Donxue Jin, Jaeyoun Jung, Ilkhyu Ha, YoungJun Choi, and Chonggun Kim. "Path hopping based on reverse AODV for security." Management of Convergence Networks and Services, pp. 574-577. Springer Berlin Heidelberg, 2006.

[11] "The Network Simulator – NS-2", available at http://www.isi.edu/nsnam/ns, 2004.

6

Energy Efficient Routing of Wireless Sensor Networks Using Virtual Backbone and life time Maximization of Nodes

Umesh B.N[1], Dr G Vasanth[2] and Dr Siddaraju[3]

[1]Research Scholar, [2]Professor & Head, Dept of Computer Science, Govt Engg College, K.R Pate and [3]Professor & Head, Dept of Computer Science, Dr. AIT, Bangalore

[1]umeshbn5@gmail.com, [2]gvasanth_ss@yahoo.co.in, [3]Siddaraju.ait@gmail.com

ABSTRACT

Different approaches have been proposed based on routing in WSN (Wireless Sensor Nodes) for Energy Efficient routing in Multi hop Networks and some localized based on geographical routing schemes have been also proposed. The key concept in virtual backbone scheduling is to minimize the energy consumption and more throughputs. To achieve QOS and fault tolerance of these backbone nodes in Multi hop Networks requires stable links. Hence existing localized routing in virtual backbone scheduling cannot guarantee the energy efficient routes. In this paper we propose an energy efficient routing for Virtual Back Bone Nodes (VBS) in which it maximizes the node life and turns off it's radio when they are in sleep mode, in order to consume less energy. A concept of Restricted Back Bone Neighborhood Routing is proposed, which assures the efficient routing with minimum energy consumption of nodes and also implemented the critical transmission radius for Backbone nodes.

Index terms: WSN, VBS, QOS, Critical Transmission radius.

I. INTRODUCTION

A wireless sensor network is a collection of nodes organized into a cooperative network. Each node consists of processing capability (one or more microcontrollers, CPUs or DSP chips), may contain multiple types of memory (program, data, and flash memories), have a RF transceiver (usually a single omni-directional antenna), have a power source (e.g., batteries and solar cells), and accommodate various sensors and actuators. The nodes communicate wirelessly and often self-organize after being deployed in an ad hoc fashion. Systems of 1000s or even 10,000 nodes are anticipated. The unique feature of Wireless Sensor Network (WSN) is the co-operative feature of sensor nodes. Wireless Sensor Networks are different from traditional network because of their own design and resource constraints and having so many challenges and issues, among these all maximizing the network life by reducing power consumption is one of the most important challenges for WSN.

Micro Electro-Mechanical systems have made extensive product developments which are of low-power and low cost sensors and less cost, such that the sensors are widely deployed. These Sensor nodes are widely used for monitoring applications like whether monitoring, agriculture monitoring etc. Conservation of energy and scalability are the two most important issues in WSN because Nodes are powered with the battery. The WSN contains three parts: Data Collection, BS (Base Station) and Management centre. Hence there is no particular infrastructure. Flooding is a kind of broadcasting in sensor network. Energy is consumed when each node retransmits the broadcasted massage which it receives. When there is an interference, node raises its energy consumption for packet retransmission which utilizes more energy, this problem is called as Broadcast Storm Problem [15][16] Extensive research has been performed on formation of back bone nodes. These backbone nodes will eliminate unwanted transmission links by turning off the redundant nodes but still backbone node assures network connectivity to deliver the data efficiently.

Subset of active nodes are called as backbone, hence backbone network is to be connected. Backbone is mainly used to improving the routing procedure which increases bandwidth efficiency, decreases overall energy consumption and increases node lifetime. Among computational functions the radio consumes more energy [1]. Hence we focus on load distribution of nodes by creating the Backbone nodes which are active when any massage sending process takes place and goes to sleep by turning off their radios. Hence it does not disturb the communication quality. However creating a single backbone node does not provide the maximum life time of node; hence it is best idea to create multiple disjoint Connected Dominated Sets (CDS) which works efficiently and can adapt to network topology changes [2].CDS can be classified into different types 1) UDG (Unit Disk Graph) and 2) DGB (Disk Graphs with Bidirectional links. UDG and DGB is an NP-hard [14] [17]. Here non CDS nodes are put into sleep mode so as to conserve overall energy of the network [11]. Backbone nodes use the deterministic scheme to keep the backbone node small with high computation [12]. Thus this scheme guarantees the CDS in connected network [13]. In previous localized routing the consumption of energy was more than it's optimal. In this paper we study energy efficient routing using virtual back bone to maximize its life time. Contributions are as follows:

- The Network is divided into zones or areas, where each backbone node present in these zones or areas has transmission radius and is restricted to find the neighbor backbone node within their transmission radius so as to find the stable links.

- A technique called Restricted Back Bone Neighborhood routing is proposed, which assures the efficient routing with minimum energy consumption. In which the backbone nodes select the neighboring backbone node which is inside the transmission radius and forms the Connected Dominated Set (CDS) of backbones from the sink node. If no node found inside the transmission radius, the backbone node extends its transmission radius called as critical transmission radius to find the next backbone node and forms a Connected Dominated Set of backbone nodes.

- A Sink Node is placed at the center which is active all the time

- Back Bone Nodes can get their lifetime using duty cycle specified to the particular zone or area.

Thus the backbone reduces communication overhead, increases bandwidth efficiency, decreases overall network energy consumption and last increases the node life span in the WSN [14]

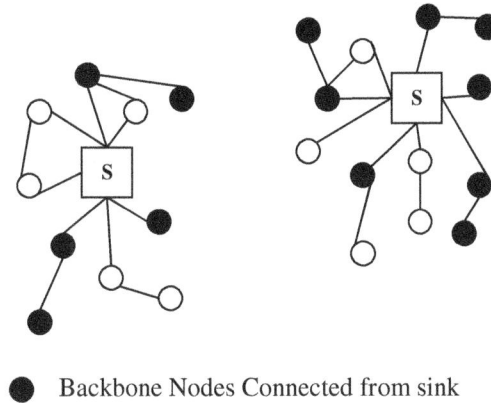

●   Backbone Nodes Connected from sink

Fig 1: Formation of CDS from Sink Node

The rest of the paper is organized as follows:
Section II presents Problem Definition, Section III presents Related Work, Section IV presents Implementation, Section V presents simulation results and Section VI presents Conclusion and Future work.

II Problem Definition

The following problem is about Wireless Sensor Network (WSN) that sensor nodes are deployed randomly in a field. Thus a battery is the main energy source for the sensor nodes. For the whole network there is an only one sink node, which is active all time and has more energy back up. Sensor nodes have links which are of identical way i,e bidirectional communication range. The radio of sensor node is the one which consumes more energy resources. Many of the algorithms select backbone node without considering any geographical information. Hence we consider formation of stable backbone links and the scheduling of radio which was proposed for static nodes previously.

III RELATED WORK

A. Network Model and Problem Definition

In Network Model, Let us consider a V as a set of n wireless devices which is widely distributed in a compact and convex region. Hence the region Ω is a unit-area square or a unit-area disk. Each node position is known by some low power receivers like GPS When single-hop broadcasting is done. For each node we can get the location information of all nodes within its transmission radius. The Transmission radius 'r' is set to all nodes uniformly. Whereas when considered about multi hop network modeled by a graph which is known as communication graph which is denoted as G (V, r), where two nodes are connected in G (V, r) if it is in Euclidean distance.

B. Localized Routing

In geometric area where multi-hop wireless networks allow the idea called localized routing protocol. The popular and widely localized protocol based on routing is greedy routing. The greedy routing is able to find the shortest paths between nodes by using local information, without global network topology.

The routing in greedy takes place when a node forwards information to the nearest destination node. Greedy routing is simple and efficient but cannot guarantee the packet delivery, while face routing can guarantee the delivery but may take a lengthy exploration. One natural improvement is to combine greedy routing and face routing by using face routing to recover the route after greedy method fails in local minimum. Many routing protocols [3], [4], [5], [6] used this approach, such as greedy face routing (GFG) [4].

Stojmenovis proposed power-aware localized routing which is based on the combination of remaining battery power in nodes and based on transmission power related to node distance [3]

Seade et al [18] proposed a power-aware greedy routing which concentrates on transmission power and also the reliability of each link. This focuses on routing without lossy links. Hence showed the significant enhancement in delivery rate and energy efficiency in lossy network

C. Energy Efficient Routing

Since energy is a scarce resource which limits the life of wireless networks, a number of energy efficient routing protocols [7], [8], [3], [4], [5], [6], [9], [10] have been proposed recently using a variety of techniques. Classical routing algorithm maybe adapted to take into account energy-related criteria rather than classical metrics such as delay or hop distance. Most of the proposed energy-aware metrics are defined as a function of the energy required to communicate on a link [3] or a function of the nodes remaining lifetime [3]. However, to minimize the global consumed energy of selected route, most of minimum energy routing algorithms are centralized algorithms. In this paper, we focus on stateless localized routing. Thus, we only review the following related work about energy efficient techniques for localized routing which address how to save energy when making local routing decision.

IV IMPLEMENTATION

In this section we describe about the backbone scheduling to find the efficient routing based on localized efficient back bone routing. The complete modules have been discussed below.

A. Schedule Transition Graph (STG)

STG is shown in the below figure, it is a centralize approximation algorithm. The time scale is shown in horizontal which is counted in rounds. The possible states of the backbone nodes are vertically listed in each round. The number of backbone is equal to the number of possible states in each round. There is an one to one mapping between state and backbone. Energy is used in 1 round which represents the time laps during each round when consumes energy. Transition of node is not allowed when there is state depletion in node.

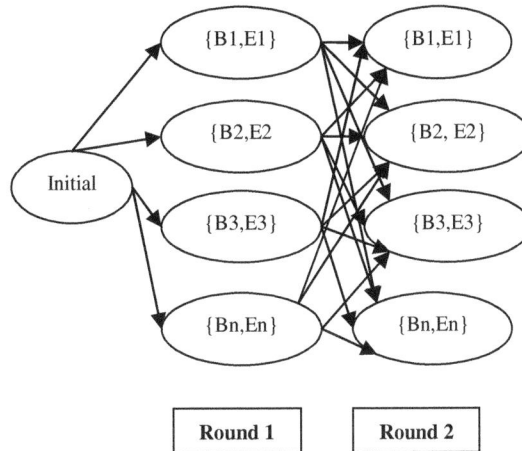

Fig 2: STG Scheduling.

Where B is a set of backbone nodes. Backbone nodes work with rounds. Connectivity of all is a network connected sub-graph and all other nodes are 1-hop away from node in B_i.

B. Localized Energy Efficient Routing

This section describes the details of our routing concept based on localized routing. This is a kind of greedy method routing. In greedy routing the node selects the next hop neighbor which is based on its distance to destination. Hence such routing may not be energy efficient. So as to achieve energy efficient, the forwarding link should have larger energy mileage. To consume less energy the distance traveled totally should be small. Thus a new concept called restricted region is introduced to restrict forwarding directions.

- The backbone node with a message which has to be forwarded finds the neighbor backbone node among the neighbors present inside the restricted area. The best neighbor has the maximum energy mileage or a stable link

- If no backbone node is present in the restricted area, it extends its critical transmission radius and finds the best node.

- If in case no node present in the area or zone then the classic greedy routing takes place.

The dataflow of this process is shown in Fig 3.

C. Critical Transmission Radius

In this study, while routing the packets from source to destination, the intermediate nodes may drop the packets before it reaches the destination, this happen because it is not able to find the better neighbor node. Hence to ensure successful and efficient routing, the backbone node should have sufficiently large transmission radius so as to find the better neighbor backbone node.

The critical transmission radius is given as

$$\rho_A(P_n) = \sqrt{\frac{\beta_A \ln (D.n)}{n\pi}}$$

ρ_A is an generalized routing method and P_n is number of nodes in network.

β_A is ratio of center of intermediate node u with radius r of the forwarding area from where the intermediate backbone node u can choose its next neighbor backbone node w

V PERFORMANCE EVALUATION

In this section we describe our simulation and Methodology as well comparing performance through simulation results of Energy model, Network lifetime of backbone nodes and Efficient Routing.

A. Simulation Methodologies

To find out the efficient routing of backbone node and network life time we simulate our proposed system using ns2 simulator.

B. Simulation Configurations

Our simulation is conducted with the Network Simulator (NS) 2.34 environment on a platform with GCC-4.3 and Ubuntu 11.10. The system is running on a laptop with Core 2 Duo T7250 CPU and 3GB RAM. In order to better compare our simulation results with other research works, we adopted the default scenario settings in NS 2.35.

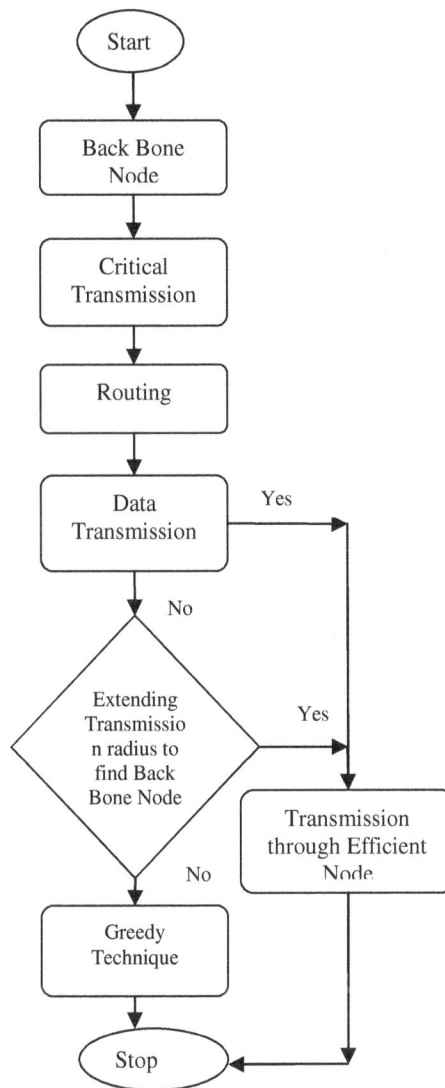

Fig 3: **Data Flow Diagram**

C. Network Life Time of Back Bone nodes

Here in this we present the result of the proposed technique. The result is shown on ns2 simulator, the concept of VBS and localized routing produces the best result in finding out the stable back bone node links which are area restricted. This is shown in Fig 8

D. Energy Model.

Energy is used only when there is a massage transfer, and turns off its radio, when a node goes to sleep mode. Energy is mainly used only when there is a message transfer which should satisfy the QOS. Consumption of less energy is shown in Fig 6

E .Efficient Routing

In this Results we represent the efficient routing, which is formed by CDS by creating the stable links between the back bone nodes. Routing efficiency is shown in Fig 7

Table 1 shows the parameter used to simulate the proposed system

Sl No	Parameter	Value
1	Number of Nodes	35
2	Topology Dimension	670m x 670m
3	Traffic Type	CBR
4	Radio Propagation Model	Two-Ray Ground
5	MAC Type	802.11 MAC Layer
6	Packet Size	512 Bytes
7	Antenna Type	Omni Dimension

Table 1: Parameters Used

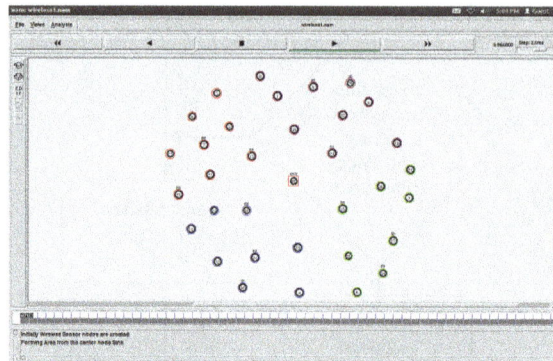

Fig 4: Initial WSN Setup with Backbone Nodes

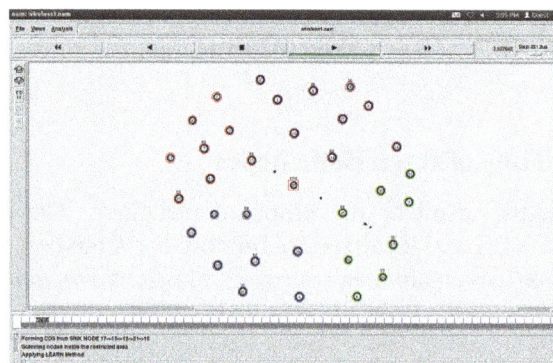

Fig 5: Routing through Backbone Nodes

SIMULATION AND ANALYSIS

I.   Energy Consumption

In WSN Energy is the major concern, hence to minimize energy consumption and for more throughput, creating virtual nodes is the idea. It is reasonable to approximatively calculate the energy consumed according to the data transmission Fig 6 shows the energy consumption of back bone nodes during the transmission of data and while in sleep mode it turns off the radios for saving energy. The graph is plotted for the no of backbone nodes present in the network and their energy consumption. Therefore the energy consumption of the proposed scheme is not higher than that of existing scheme. The blue line shows the energy consumed by the backbone nodes in the overall WSN

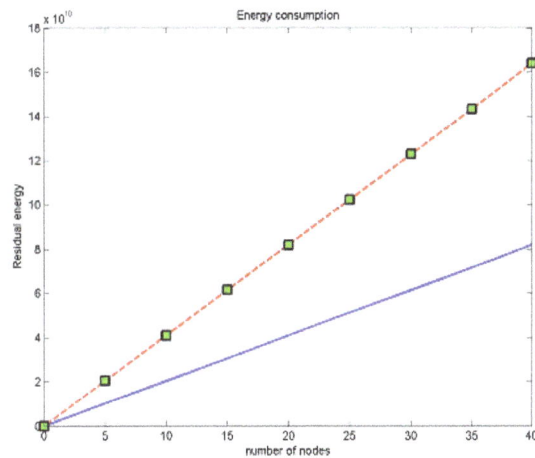

Fig 6: Energy Consumption

II.   Efficient Routing

In efficient routing the total length of the path found by LEARN is within a constant optimum. The proposed system is compared with existing localized routing methods and proves that it can guarantee the energy efficient routes from source to destination. Extensive simulation is conducted to study the performance of the LEARN Routing. This shows that LEARN Localized routing method guarantees the energy efficient routes in random network with high probability. Fig 7 shows the backbone nodes with efficient routing. The simulation results for throughput is shown as efficient routing

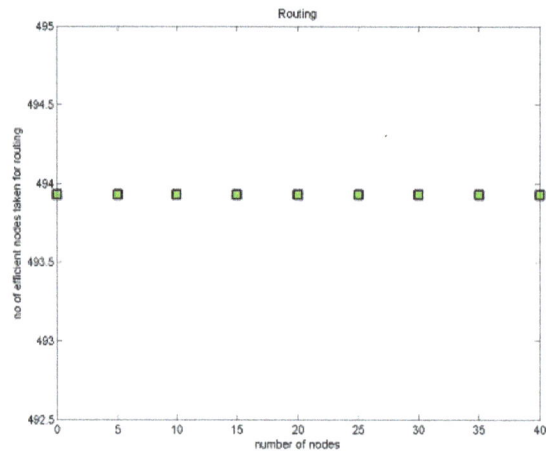

Fig 7: Efficient Routing

III.   Node Lifetime

In node lifetime backbone nodes achieves the prolong network life. Identical initial energy and the imbalanced energies are used as energy configurations. Initially the transmission range is fixed, later based on the critical transmission the transmission radius can be extended. Fig 8 shows the achieved network lifetime. By distributing the energy uniformly. The blue line indicates the network lifetime of backbone nodes

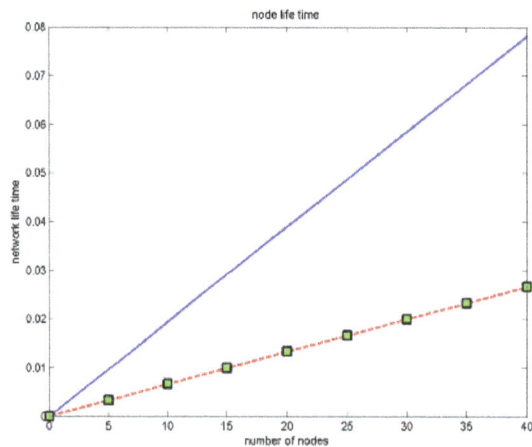

Fig 8: Node Lifetime

IV.   Delay

This simulation results shows that there is a less delay in finding out efficient path and constructing CDS from the sink node it is shown in Fig 9

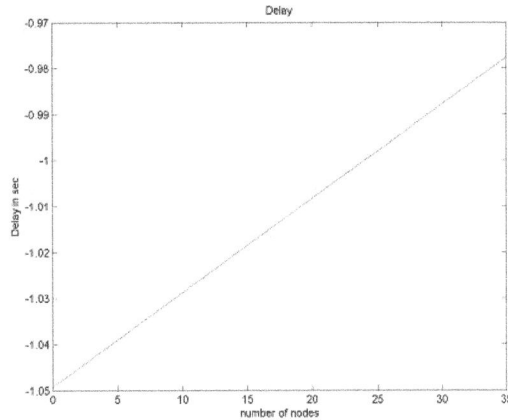

Fig 9: Delay

VI Conclusion

Energy Efficient communication is needed for WSN so as to work for a longer period because of battery powered, therefore network lifetime through power aware node organization is high desirable. An efficient method based on low energy consumption scheduling has to be implemented so as to schedule the node activity between sleep and active state. One solution is to create a backbone node and make their radios off when they are in sleep mode and the routing should be energy efficient by making the network life time prolong.

REFERENCES

[1] Shnayder, M. Hempstead, B.-r. Chen, G. W. Allen, and M. Welsh, "Simulating the power consumption of large-scale sensor network applications," in Proc. of SenSys '04. New York, NY, USA: ACM, 2004, pp. 188–200.

[2] C. Misra, R. Mandal, "Rotation of cds via connected domatic partition in ad hoc sensor networks," IEEE Trans. on Mobile Computing, pp. 488–499, 2009.

[3] I. Stojmenovic and X. Lin, "Loop-Free Hybrid Single-Path/ Flooding Routing Algorithms with Guaranteed Delivery for Wireless Networks," IEEE Trans. Parallel and Distributed Systems, vol. 12, no. 10, pp. 1023-1032, Oct. 2001.

[4] P. Bose, P. Morin, I. Stojmenovic, and J. Urrutia, "Routing with Guaranteed Delivery in Ad Hoc Wireless Networks," ACM/ Kluwer Wireless Networks, vol. 7, no. 6, pp. 609-616, 2001.

[5] F. Kuhn, R. Wattenhofer, Y. Zhang, and A. Zollinger, "Geometric Ad-Hoc Routing: Of Theory and Practice," Proc. ACM Symp. Principles of Distributed Computing (PODC), 2003.

[6] F. Kuhn, R. Wattenhofer, and A. Zollinger, "Worst-Case Optimal and Average-Case Efficient Geometric Adhoc Routing, ," Proc. ACM Int'l Symp. Mobile Ad Hoc Networking and Computing (MobiHoc), 2003.

[7] K. Kar, M. Kodialam, T. Lakshman, and L. Tassiulas, "Routing for Network Capacity Maximization in Energy-Constrained Ad-Hoc Networks," Proc. IEEE INFOCOM, 2003.

[8] J.-H. Chang and L. Tassiulas, "Energy Conserving Routing in Wireless Ad-Hoc Networks," Proc. IEEE INFOCOM, 2000.

International Journal of Wireless & Mobile Networks (IJWMN) Vol. 5, No. 1, February 2013

[9] Q. Li, J. Aslam, and D. Rus, "Online Power-Aware Routing in Wireless Ad-Hoc Networks," Proc. ACM Mobicom, 2001.

[10] J. Kuruvila, A. Nayak, and I. Stojmenovic, "Progress and Location Based Localized Power Aware Routing for Ad Hoc and Sensor Wireless Networks," Int'l J. Distributed Sensor Networks, vol. 2, pp. 147-159, 2006.

[11] B. Chen, K. Jamieson, H. Balakrishnan, and R. Morris, "SPAN: An energy-efficient coordination algorithm for topology maintenance in ad hoc wireless networks," Wireless Networks, ACM, vol. 8, no. 5, pp. 481–494, 2002.

[12] F. Dai and J. Wu, "On constructing k-connected k-dominating set in wireless ad hoc and sensor networks," Journal of Parallel Distributed Computing (JPDC), Elsevier, vol. 66, no. 7, pp. 947–958, 2006.

[13] An extended localized algorithm for connected dominating set formation in ad hoc wireless networks," IEEE Transactions on Parallel Distributed Systems, vol. 15, no. 10, pp. 908–920, 2004.

[14] M. Rai, Sh. Verma, and Sh. Tapaswi, "A Power Aware Minimum Connected Dominating Set for Wireless Sensor Networks," Journal of networks, Vol. 4, no. 6, August 2009.

[15] J. Akbari Torkestani, M. R. Meybodi, "An intelligent backbone formation algorithm for wireless ad networks based on distributed learning automata," Computer Networks 54, pp. 826–843, 2010.

[16] S. Hussain, M. I. Shafique, and L. T. Yang, "Constructing a CDS-Based Network Backbone for Energy Efficiency in Industrial Wireless Sensor Network," In Proceedings of HPCC, pp. 322-328, 2010.

[17] Z. Liu, B. Wang, and Q. Tang, "Approximation Two Independent Sets Based Connected Dominating Set Construction Algorithm for Wireless Sensor Networks," Inform. Technol. J., Vol. 9, Issue 5, pp. 864-876, 2010.

[18] K. Seada, M. Zuniga, A. Helmy, and B. Krishnamachari, "Energy- Efficient Forwarding Strategies for Geographic Routing in Lossy Wireless Sensor Networks," Proc. ACM Int'l Conf. Embedded Networked Sensor Systems (Sensys), 2004.

7

Performance Comparison of Minimum Hop and Minimum Edge Based Multicast Routing Under Different Mobility Models for Mobile Ad hoc Networks

Natarajan Meghanathan

Jackson State University, 1400 Lynch St, Jackson, MS, USA
natarajan.meghanathan@jsums.edu

ABSTRACT

The high-level contribution of this paper is to establish benchmarks for the minimum hop count per source-receiver path and the minimum number of edges per tree for multicast routing in mobile ad hoc networks (MANETs) under different mobility models. In this pursuit, we explore the tradeoffs between these two routing strategies with respect to hop count, number of edges and lifetime per multicast tree with respect to the Random Waypoint, City Section and Manhattan mobility models. We employ the Breadth First Search algorithm and the Minimum Steiner Tree heuristic for determining a sequence of minimum hop and minimum edge trees respectively. While both the minimum hop and minimum edge trees exist for a relatively longer time under the Manhattan mobility model; the number of edges per tree and the hop count per source-receiver path are relatively low under the Random Waypoint model. For all the three mobility models, the minimum edge trees have a longer lifetime compared to the minimum hop trees and the difference in lifetime increases with increase in network density and/or the multicast group size. Multicast trees determined under the City Section model incur fewer edges and lower hop count compared to the Manhattan mobility model.

KEYWORDS

Minimum Hop, Minimum Edge, Multicast Routing, Mobile Ad hoc Networks, Simulations, Steiner Trees, Mobility Models, Tree Lifetime

1. INTRODUCTION

A mobile ad hoc network (MANET) is a dynamic distributed system of wireless nodes that move independent of each other in an autonomous fashion. The network bandwidth is limited and the medium is shared. As a result, transmissions are prone to interference and collisions. The battery power of the nodes is constrained and hence nodes operate with a limited transmission range, often leading to multi-hop routes between any pair of nodes in the network. Communication structures (e.g.., paths, trees, connected dominating sets and etc) for routing in wireless ad hoc networks could be principally based on two different approaches [1]: Optimum Routing Approach (ORA) and Least Overhead Routing Approach (LORA). With ORA, the communication structure used at any time instant is always the optimum with respect to a particular metric. On the other hand, with LORA, a communication structure determined for optimality with respect to a particular metric at a time instant is used in the subsequent time instants as long as the communication structure exists. For dynamically changing, distributed, resource-constrained MANETs, the LORA strategy is often preferred over the ORA strategy to avoid the communication overhead incurred in determining the optimum communication structure at every time instant. Hence, we focus on using LORA for the rest of this paper.

Multicasting in ad hoc wireless networks has numerous applications in collaborative and distributed computing like civilian operations (audio/ video conferencing, corporate communications, distance learning, outdoor entertainment activities), emergency search-and-rescue, law enforcement and warfare situations, where establishing and maintaining a communication infrastructure may be expensive or difficult. A common feature among all these applications is one-to-many and many-to-many communications among the participants [2].

Several MANET multicast routing protocols have been proposed in the literature [1]. They are mainly classified as: tree-based and mesh-based protocols. In tree-based protocols, only one route exists between a source and a destination and hence these protocols are efficient in terms of the number of link transmissions. The tree-based protocols can be further divided into two types: source tree-based and shared tree-based. In source tree-based multicast protocols, the tree is rooted at the source. In shared tree-based multicast protocols, the tree is rooted at a core node and all communication between the multicast source and the receiver nodes is through the core node. Even though shared tree-based protocols are more scalable with respect to the number of sources, these protocols suffer under a single point of failure, the core node. On the other hand, source tree-based protocols are more efficient in terms of traffic distribution. In mesh-based protocols, multiple routes exist between a source and each of the receivers of the multicast group. A receiver node receives several copies of the data packets, one copy through each of the multiple paths. Mesh-based protocols provide robustness at the expense of a larger number of link transmissions leading to inefficient bandwidth usage. Considering all the pros and cons of these different classes of multicast routing in MANETs, we feel the source tree-based routing protocols are more efficient in terms of traffic distribution and link usage. Hence, all of our work in this research will be in the category of on-demand source tree-based multicast routing.

Not much work has been done towards the evaluation of MANET multicast routing from a theoretical point of view with respect to metrics such as the hop count per source-receiver path and the number of edges per multicast tree and their impact on the lifetime per multicast tree. These two theoretical metrics significantly contribute and influence the more practically measured performance metrics such as the energy consumption per node, end-to-end delay per data packet, multicast routing overhead and etc. that have been often used to evaluate and compare the different MANET multicast routing protocols in the literature. Hence, we take a different approach in this paper. We study MANET multicast routing using the theoretical algorithms that would yield the benchmarks (i.e., optimum values) for the above two metrics – the Breadth First Search (BFS) algorithm [3] for the minimum hop count per source-receiver path and the minimum Steiner tree heuristic [4] for the minimum number of edges.

Our simulation methodology is outlined as follows: Using the mobility profiles of the nodes gathered offline from a discrete-event simulator (ns-2 [6]), we will generate snapshots of the MANET topology, referred to as Static Graphs, periodically for every fixed time instant. For simulations with a particular algorithm, if a multicast tree is not known for a particular time instant, we will run the algorithm on the static graph in a centralized fashion and adopt the LORA strategy of using this multicast tree as long as it exists for the subsequent static graphs. If the tree no longer exists after a certain time instant, the multicast algorithm is again run to determine a new tree. This procedure is repeated for the entire simulation time. Depending on the algorithm used, the sequence of multicast trees generated either have the minimum hop count per source-receiver path or the minimum number of edges. Our hypothesis is that the multicast trees, determined to optimize one of the two theoretical metrics, would be sub-optimal with respect to the other metric. Through extensive simulations, we confirm our hypothesis to be true and we explain in detail the performance tradeoffs associated with the two metrics.

The rest of the paper is organized as follows: Section 2 reviews the existing related work in the literature. Section 3 introduces the notion of a Static Graph and reviews the BFS algorithm for minimum hop path trees and Kou et al.'s heuristic for minimum edge Steiner trees. Section 4 briefly describes the three mobility models (Random Waypoint, City Section and Manhattan models) simulated in this paper. Section 5 presents the simulation results for the benchmark values of the two theoretical metrics, explores the tradeoffs between these metrics and their impact on the lifetime per multicast tree under each of the three mobility models. Section 6 concludes the paper. For the rest of the paper, the terms 'vertex' and 'node', 'algorithm' and 'heuristic', 'destination' and 'receiver' are used interchangeably. They mean the same.

2. REVIEW OF RELATED WORK IN THE LITERATURE

Several MANET multicast routing protocols have been proposed in the literature [1][2]. They are mainly classified as: tree-based and mesh-based protocols. In tree-based protocols, only one route exists between a source and a destination and hence these protocols are efficient in terms of the number of link transmissions. The tree-based protocols can be source tree-based or shared tree-based. In source tree-based protocols, the tree is rooted at the source. In shared tree-based protocols, the tree is rooted at a core node and all communication between the multicast source and the receiver nodes is through the core node. Even though shared tree-based multicast protocols are more scalable with respect to the number of sources, these protocols suffer under a single point of failure, the core node. On the other hand, source tree-based protocols are more efficient in terms of traffic distribution. In mesh-based multicast protocols, multiple routes exist between a source and each of the receivers of the multicast group. A receiver node receives several copies of the data packets, one copy through each of the multiple paths. Mesh-based protocols provide robustness at the expense of a larger number of link transmissions leading to inefficient bandwidth usage. Considering all the pros and cons of these different classes of multicast routing in MANETs, we feel the source tree-based multicast routing protocols are more efficient in terms of traffic distribution and link usage. Hence, all of our work in this research will be in the category of on-demand source tree-based multicast routing.

Some of the recent performance comparison studies on MANET multicast routing protocols reported in the literature are as follows: In [11], the authors compare the performance of the tree-based MAODV and mesh-based ODMRP protocols with respect to the packet delivery ratio and latency. In [12], the authors propose a stability-based multicast mesh protocol and compare its performance with ODMRP. [13], the authors compare a dominating set-induced mesh based multicast routing protocol for efficient flooding and control overhead and compare the protocol's performance with that of MAODV and ODMRP. In [14], the authors explore the use of genetic algorithms to optimize the performance the performance of tree and mesh based MANET multicast protocols with respect to packet delivery and control overhead. The impact of route selection metrics such as hop count and link lifetime on the performance of on-demand mesh-based multicast ad hoc routing protocols has been examined in [15]. In [16], the author has proposed non-receiver aware and receiver-aware (depending on whether the nodes in the network are aware of the multicast group or not) extensions to the Location Prediction Based Routing (LPBR) protocol to simultaneously minimize the edge count, hop count and number of multicast tree discoveries. An agent-based multicast routing scheme (ABMRS) that uses a set of static and mobile agents for network and multicast initiation and management has been proposed in [17] and compared with MAODV. A zone-based scalable and robust location aware multicast algorithm (SRLAMA) has also been recently proposed for MANETs [18].

3. REVIEW OF THE GRAPH THEORY ALGORITHMS

In this section, we first describe the notion of a static graph, referring to the snapshots of the network topology, on which we run the theoretical algorithms to simulate multicasting. We then describe the two algorithms (BFS and Minimum Steiner tree heuristic) used in this paper.

3.1. Static Graph

A static graph is a snapshot of the MANET topology at a particular time instant. Using the mobility profiles of the nodes generated offline from ns-2, we will be able to determine the locations of a node at any particular time instant. A static graph $G(t) = (V, E)$ generated for a particular time instant t, comprises of all the nodes in the network as the vertex set V; there exists an edge $(u, v) \in E$, if and only if, the Euclidean distance between the two end vertices u and $v \in V$, is less than or equal to the transmission range of the nodes in the network. All the edges in E are of unit weight. We assume a homogeneous network of nodes and all nodes operate at an identical and fixed transmission range.

3.2. Breadth First Search (BFS)

The BFS algorithm has been traditionally used to check the connectivity of a network graph. When we start the BFS algorithm on a randomly chosen node, we should be able to visit all the vertices in the graph, if the graph is connected. BFS returns a tree rooted at the chosen start node; when we visit a vertex v for the first time in our BFS algorithm, the vertex u through which we visit v is considered as the predecessor node of v in the tree. Every vertex in the BFS tree, other than the root node, has exactly one predecessor node. When we run BFS on a static graph with unit edge weights, we will be basically obtaining a minimum hop multicast tree such that every node in the graph is connected to the root node (the source node of the multicast group) of the tree on a path with the theoretically minimum hop count.

Figure 1 illustrates BFS in the form of a pseudo code and Figure 2 demonstrates the step-by-step execution of BFS on a sample graph. If $MG \subseteq V$ represents the multicast group – set of receiver nodes and a source node s, we start BFS at s and visit all the vertices in the network graph. Once we obtain a BFS tree rooted at s, we trace back from every receiver $d \in MG$ and determine the minimum hop s-d path. The minimum hop multicast tree is an aggregate of all these minimum hop paths connecting the source s to receiver d in the multicast group.

The set of vertices represented in parentheses below each of the graphs in Figure 2 represents the FIFO-Queue data structure used to maintain the list of vertices that are visited but whose neighbours are yet to be explored (refer the pseudo code in Figure 1). The vertices stored in this queue are extracted in a First-In First-Out fashion and their neighbours are visited if they have not been already explored. Note that, for simplicity, we restrict our research in this paper to only single source multicast groups; the research could be easily extended for multicast groups with more than one source node. Once we establish the benchmarks for single source multicast groups in this paper, we will extend the research for multi-source multicast groups in the immediate future.

Input: Static Graph $G = (V, E)$, source s
Auxiliary Variables/Initialization: Nodes-Explored = Φ, FIFO-Queue = Φ, root-node
$$\forall\ v \in V,\ \text{Predecessor}(v) = \text{NULL}$$
Begin Algorithm *BFS* (G, s)
   root-node = s
   Nodes-Explored = Nodes-Explored U {root-node}
   FIFO-Queue = FIFO-Queue U {root-node}
   while (|FIFO-Queue| > 0) **do**
      first-node u = Dequeue(FIFO-Queue) // extract the first node
      for (every edge $(u, v) \in E$) **do** // i.e. every neighbour v of node u
         if ($v \notin$ Nodes-Explored) **then**
            Nodes-Explored = Nodes-Explored U {v}

```
        FIFO-Queue = FIFO-Queue U {v}
        Predecessor (v) = u
    end if
  end for
end while
```

End Algorithm *BFS*

Figure 1: Pseudo Code for Breadth First Search (BFS)

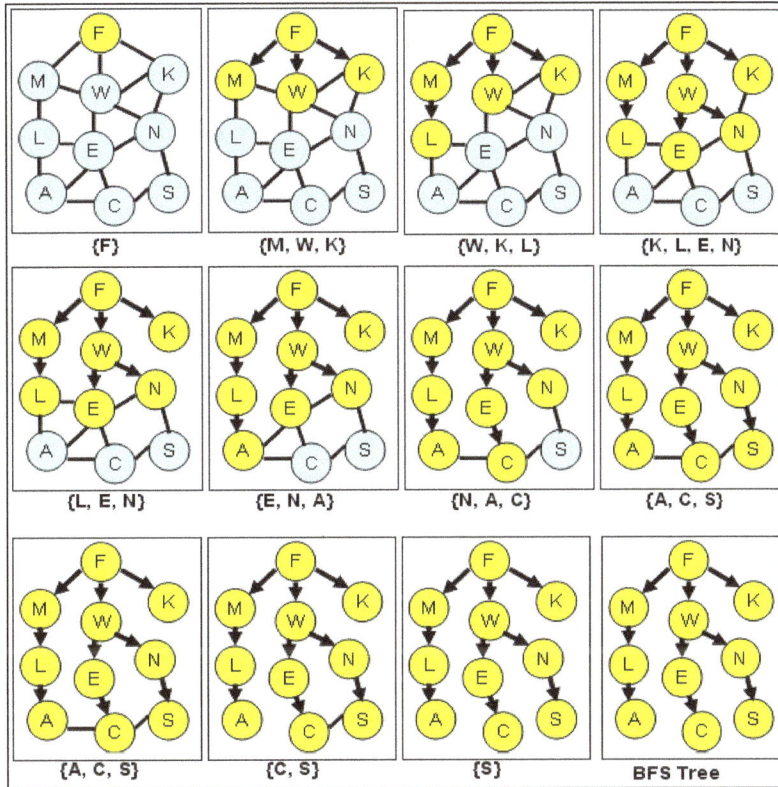

Figure 2: Execution of BFS on a Sample Graph

3.3. Minimum Edge Multicast Steiner Tree

Given a static graph, $G = (V, E)$, where V is the set of vertices, E is the set of edges and a subset of vertices (called the multicast group or Steiner points) $MG \subseteq V$, the multicast Steiner tree is the tree with the least number of edges required to connect all the vertices in MG. Unfortunately, the problem of determining a minimum edge Steiner tree in an undirected graph like that of the static graph is NP-complete. Efficient heuristics (e.g., [4]) have been proposed in the literature to approximate a minimum Steiner tree. In this paper, we use the Kou et al's [4] well-known $O(|V||MG|^2)$ heuristic ($|V|$ is the number of nodes in the network graph and $|MG|$ is the size of the multicast group comprising of the source nodes and the receiver nodes) to approximate the minimum edge Steiner tree in graphs representing snapshots of the network topology. An *MG-Steiner-tree* is referred to as the minimum edge Steiner tree connecting the set of nodes in the multicast group $MG \subseteq V$. In unit disk graphs such as the static graphs used in our research, Step 5 of the heuristic is not needed and the minimal spanning tree T_{MG} obtained at the end of Step 4 could be considered as the minimum edge Steiner tree.

Input: A Static Graph $G = (V, E)$
Multicast Group $MG \subseteq V$
Output: A *MG-Steiner-tree* for the set $MG \subseteq V$

Begin Kou et al Heuristic (G, MG)
  Step 1: Construct a complete undirected weighted graph $G_C = (MG, E_C)$ from G and MG where $\forall (v_i, v_j) \in E_C$, v_i and v_j are in MG, and the weight of edge (v_i, v_j) is the length of the shortest path from v_i to v_j in G.
  Step 2: Find the minimum weight spanning tree T_C in G_C (If more than one minimal spanning tree exists, pick an arbitrary one).
  Step 3: Construct the sub graph G_{MG} of G, by replacing each edge in T_C with the corresponding shortest path from G (If there is more than one shortest path between two given vertices, pick an arbitrary one).
  Step 4: Find the minimal spanning tree T_{MG} in G_{MG} (If more than one minimal spanning tree exists, pick an arbitrary one). Note that each edge in G_{MG} has weight 1.

  return T_{MG} as the *MG-Steiner-tree*

End Kou et al Heuristic

Figure 3: Kou et al's Heuristic [4] to find an Approximate Minimum Edge Steiner Tree

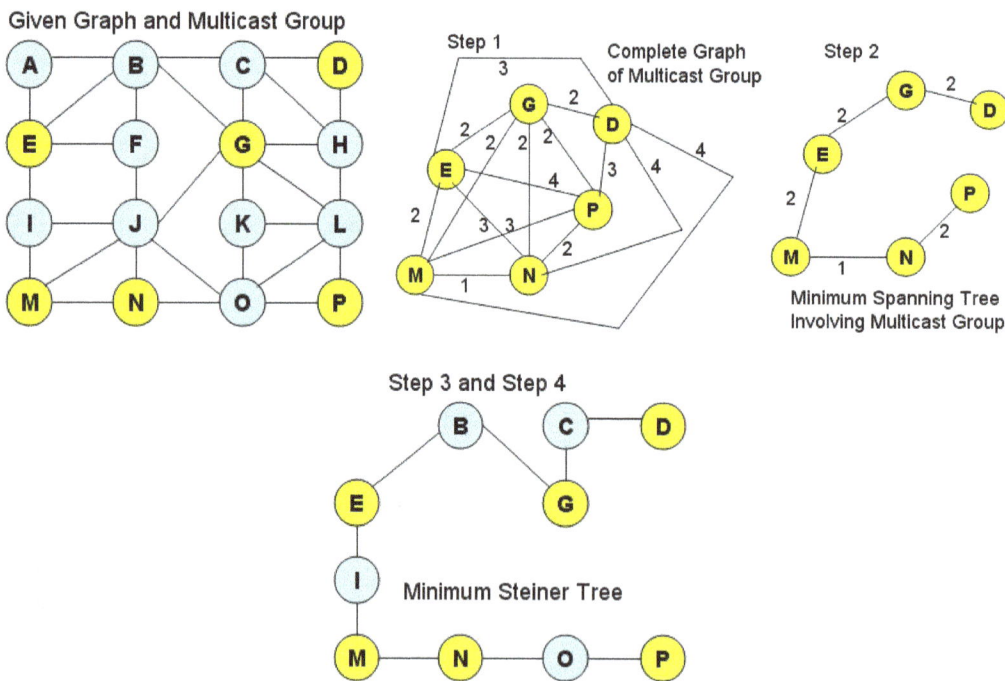

Figure 4: Example to Illustrate the Construction of a Minimum Steiner Tree

We give a brief outline of the heuristic in Figure 3 and illustrate the working of the heuristic through an example in Figure 4. The vertices {D, G, E, M, N, P} form the multicast group in the vertex set {A, B … P}. As observed in the example, the subgraph G_{MG} obtained in Step 3 is nothing but the minimal spanning tree T_{MG}, which is the output of Step 4. In general, for unit disk graphs, like the static graphs we are working with, the outputs of both Steps 3 and 4 are the same and it is enough that we stop at Step 3 and output the MG-Steiner-tree.

4. REVIEW OF THE MOBILITY MODELS

In this section, we provide a brief overview of the Random Waypoint mobility model [8] commonly used in MANET simulation studies and the widely used mobility models for vehicular ad hoc networks (VANETs), viz., City Section [9] and Manhattan mobility models [10]. All the three mobility models assume the network to be confined within fixed boundary conditions. The mobility of a node is independent of the other nodes in all the three mobility models. Under the Random Waypoint model, each node can move anywhere within a network region. For the two VANET models, the network is assumed to be divided into grids of square blocks with identical block length. The network for the City Section and Manhattan models is thus basically composed of a number of horizontal and vertical streets with each street having two lanes, one for each direction (east and west direction for horizontal streets; north and south direction for vertical streets); nodes can move only along the grids of horizontal and vertical streets. All streets are assumed to have identical value for the maximum speed limit (v_{max}).

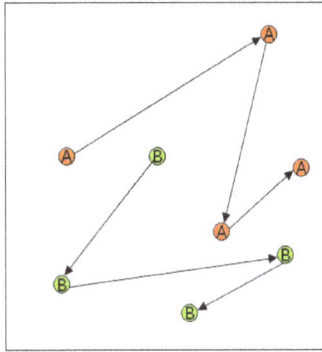

Figure 5: Random Waypoint Mobility Model

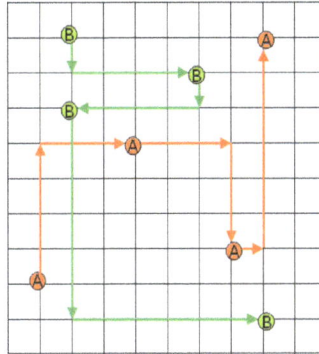

Figure 6: City Section Mobility Model

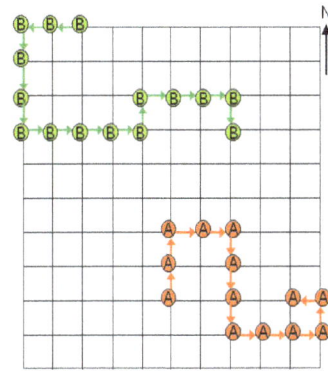

Figure 7: Manhattan Mobility Model

4.1. Random Waypoint Mobility Model

The nodes are initially assumed to be placed at random locations in the network. As each node moves independently of the other nodes in the network, the mobility pattern described here is applicable for every node. The movement of a node is described as follows: The node randomly chooses a target location (within the network) to move. The velocity to move to the chosen location is uniform randomly chosen from the interval $[v_{min},...,v_{max}]$. The node is assumed to move on a straight line to the chosen location with the chosen velocity. After reaching the targeted location, the node may stay there for a certain time, called the pause time, and then continues to move by choosing a different target location (that is independent of the current and previous locations) under a different randomly chosen velocity from the interval $[v_{min},...,v_{max}]$. Each time a node changes direction, it is referred to as moving to a new waypoint. Figure 5 illustrates the mobility of two nodes (A and B) moving in random directions with randomly chosen velocities anywhere within a network. In our simulations in this paper, the values of both v_{min} and the pause time are 0.

4.2. City Section Mobility Model

To start with, each node is placed at a randomly chosen street intersection. Note that it is possible for more than one node to be placed at a particular street intersection. The mobility of a node is described as follows: Each node chooses a random street intersection (within the grid network) to move with a velocity uniform-randomly chosen from the range [0, …, v_{max}]. The node then moves to the chosen street intersection with the chosen velocity on a path that will incur the least amount of travel time. Any tie, between two or more paths that offer the same minimum travel time, is broken arbitrarily. After moving to the chosen street intersection, the node may stay there for a pause time (in our simulations, there is zero pause time) and then continues to move by randomly choosing a different target street intersection (independent of the current and previous street intersections) under a different uniform-randomly chosen velocity form the range [0, …, v_{max}]. The above procedure is independently repeated by each node. Figure 6 illustrates the movement of two nodes (A and B) under the City Section model.

4.3. Manhattan Mobility Model

The nodes are initially assumed to be placed in randomly chosen street intersections. The mobility of a node is decided one street block at a time. In Figure 7, to start with, node A has equal chance (25% each) to move in each of the four possible directions (east, west, north or south) starting from its initial location; whereas, node B can move only either to the west, east or south with a 1/3 chance for each direction. The velocity at which a node moves from one street to the subsequent street intersection is uniform-randomly chosen from the range [0, …, v_{max}]. After a node moves to the chosen neighbouring street intersection, the subsequent street intersection to which the node will move is chosen probabilistically. If a node can continue to move in the same direction or can also change directions, the node has 50% chance of moving in the same direction; 25% chance to turn on either side, with the exact new direction depending on the direction of the previous movement. If a node has only two options, then the node either moves to the next street intersection by continuing in the same direction or changes direction. For example, in Figure 7, after node A reaches the rightmost network boundary, it can either move to the north or to the south, each with a probability of 0.5 and the node chooses to move in the north direction. After moving to the next street intersection, node A can continue to move northwards or turn left and move eastwards, each with a probability of 0.5. If a node has only one option to move (this situation occurs when a node reaches any of the four corners forming the network boundary), then the node has no other choice except to explore that option. For example, in Figure 7, the only option for node B, which was initially travelling westward and reaching the corner of the network, is to turn to the left and proceed southwards.

5. SIMULATIONS

The simulations have been conducted in a discrete-event simulator implemented by the author in Java. The two multicast algorithms have been implemented in a centralized fashion. We generate the static graphs by taking snapshots of the network topology, periodically for every 0.25 seconds, and run the two multicast algorithms. The simulation time is 1000 seconds. We consider a square network of dimensions 1000m x 1000m. The transmission range of the nodes is 250m. The network density is varied by performing the simulations with 50 nodes (low density) and 100 nodes (high density). We assume there is only one source for the multicast group and three different values for the number of receivers per multicast group are considered: 3 (small), 10 (moderate) and 18 (large). A multicast group comprises of a source node and a list of receiver nodes, the size of which is mentioned above. The v_{max} values used for each of the three mobility models (Random Waypoint, City Section and Manhattan models) are 5 m/s (low mobility), 25 m/s (moderate mobility) and 50 m/s (high mobility). The pause time is 0 seconds. Section 4 provides a detailed description on the behaviour of the mobility models.

The performance metrics measured are as follows. Each performance metric illustrated in Figures 8 through 17 is measured using 5 different lists of receiver nodes for the same size and the multicast algorithm is run on five different mobility trace files generated for a particular value of v_{max} for each mobility model:

(i) *Tree Connectivity*: This metric refers to the percentage of time instants there exists a multicast tree connecting the source node to the receiver nodes of the multicast group, averaged over the mobility profiles generated for a particular value of v_{max} for a given number of network nodes and number of receivers per multicast group.

(ii) *Number of Links per Tree*: This metric refers to the total number of links in the entire multicast tree, time-averaged over the duration of the multicast session. For example, a multicast session uses two trees, one tree with 10 links for 3 seconds and another tree with 15 links for 6 seconds, then the time-averaged value for the number of links per tree for the 9-second duration of the multicast session is (10*3 + 15*6)/(3 + 6) = 13.3 and not 12.5.

(iii) Number of Hops per Receiver: We measure the number of hops in the paths from the source to each receiver of the multicast group and average it for the duration of the multicast session. This metric is also a time-averaged value of the number of hops from a multicast source to a receiver and then averaged over all the receivers of a multicast session.

(iv) *Lifetime per Multicast Tree*: Whenever a link break occurs in a multicast tree, we establish a new multicast tree. The lifetime per multicast tree is the average of the time between successive multicast tree discoveries for a particular routing protocol or algorithm, over the duration of the multicast session. The larger the value of the lifetime per multicast tree, the lower the number of multicast tree transitions or discoveries needed.

5.1. Tree Connectivity

The connectivity of the trees (refer Figure 8) does not depend on any individual multicast algorithm used and depends only on the mobility model, network density, node mobility and the number of receivers per multicast group. The Manhattan model incurs the lowest tree connectivity for most of the scenarios, especially for those with low network density (number of nodes) and larger multicast group size. On the other hand, the Random Waypoint model incurs the largest tree connectivity.

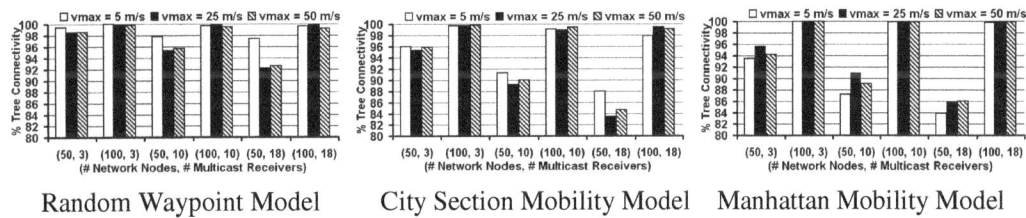

Random Waypoint Model   City Section Mobility Model   Manhattan Mobility Model

Figure 8: Percentage Tree Connectivity under the Different Mobility Models

For a fixed density and node mobility, as we increase the number of receivers per multicast group, the number of time instants for which we could connect the source node to all the receiver nodes decreases. With node mobility, the source may not be connected all the time to all the receivers. The probability of the source connected to all the receiver nodes decreases with increase in the number of receivers per multicast group. On the other hand, for a fixed node mobility and number of receivers per multicast group, the connectivity of a multicast tree increases with increase in the network density. This could be attributed to the availability of a larger number of nodes to connect the source node to the multicast receivers. For low density networks, we observe that as the number of receivers per multicast group increases, the percentage of tree connectivity decreases with increase in maximum node velocity. This can be attributed to an appreciable probability (in low density networks) of not being able to find a path that connects a source node to all the receiver nodes of the multicast group. As the network density increases, there are relatively less variations in tree connectivity with respect to increase in the number of multicast receivers as well as with increase in maximum node velocity.

5.2. Number of Edges per Multicast Tree and Hop Count per Source-Receiver Path

As expected, the minimum-edge based Steiner trees incurred the smallest number of edges per multicast trees. In most of the scenarios, the number of edges per multicast tree under a Random Waypoint model is larger than that incurred with the City Section model, which is larger than that incurred with the Manhattan model. On average, the number of edges per minimum hop tree is 13-35% more than those incurred with the minimum edge tree. With an objective to optimize the hop count, minimum hop based multicast trees select edges that could constitute a minimum hop path, but with a higher probability of failure in the immediate future. The physical distance between the constituent nodes of an edge on a minimum hop path is close to the transmission range of the nodes at the time of tree formation itself. For a given network density, as we increase the number of receivers per multicast group from 3 to 18, the average number of edges per multicast tree increased by a factor of 3 to 4. For the minimum hop and minimum edge trees, for a given level of node mobility and number of receivers per group, as we increase the network density, the number of edges per tree remains the same or only slightly decreases.

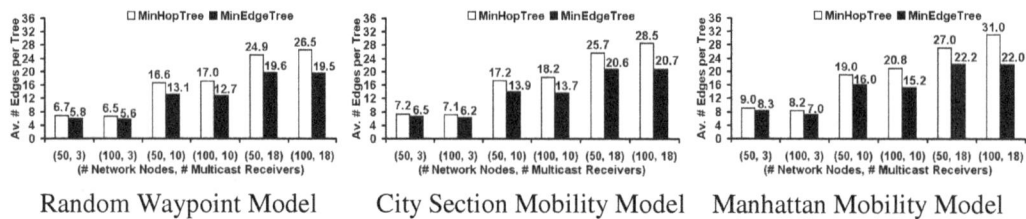

Random Waypoint Model   City Section Mobility Model   Manhattan Mobility Model

Figure 9: # Edges per Tree under Different Mobility Models (Max. Node Velocity: 5 m/s)

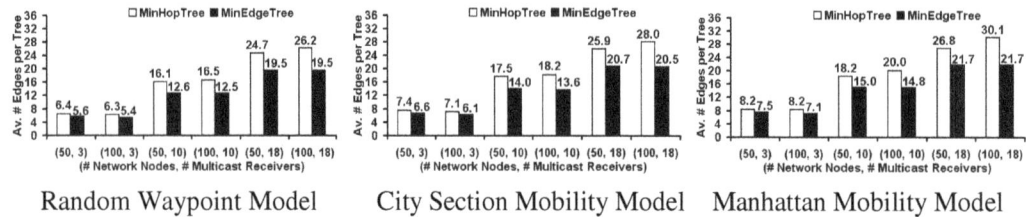

Random Waypoint Model   City Section Mobility Model   Manhattan Mobility Model

Figure 10: # Edges per Tree under Different Mobility Models (Max. Node Velocity: 25 m/s)

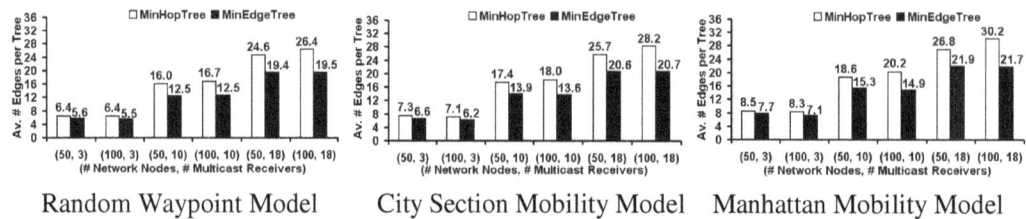

Random Waypoint Model   City Section Mobility Model   Manhattan Mobility Model

Figure 11: # Edges per Tree under Different Mobility Models (Max. Node Velocity: 50 m/s)

As expected, the minimum hop multicast trees incurred the lowest hop count per source-receiver path. In most of the scenarios, the hop count per source-receiver path for both the multicast trees incurred under a Random Waypoint model is larger than that incurred with the City Section model, which is larger than that incurred with the Manhattan model. The larger hop count per source-receiver path for minimum edge trees could be attributed to a relatively lower number of edges compared to the minimum hop trees. As we connect the source node to the multicast receivers with the lowest possible number of edges, the number of hops between the source node and to each of the receiver nodes increases. This is the tradeoff between the objectives of minimizing the number of edges per multicast tree and the hop count per individual source-receiver paths in the multicast tree.

For both minimum hop and minimum edge multicast trees, for a given network density and number of receivers per multicast group, there is appreciably no impact of the maximum node velocity on the average number of edges per tree as well as the hop count per source-receiver path. For a given level of node mobility (i.e., maximum node velocity) and network density, as we increase the number of receivers per multicast group, the average hop count per source-receiver path for minimum hop trees decreases. On the other hand, the average hop count per source-receiver path for minimum edge trees increases. This could be attributed to the relatively fewer number of edges in the minimum edge trees compared to those incurred by the minimum hop trees. The relatively more edges in minimum hop trees at larger number of receivers per multicast group results in lower hops count per source-receiver path. The average number of edges per minimum hop tree for a network of 50 nodes and 3 receivers per multicast group is about 1 edge more than those incurred by the minimum edge trees; on the other hand, the average number of edges per minimum hop tree for a network of 50 nodes and 18 receivers per multicast group is about 7 edges more than the minimum. Similar observations could be made for network of 100 nodes.

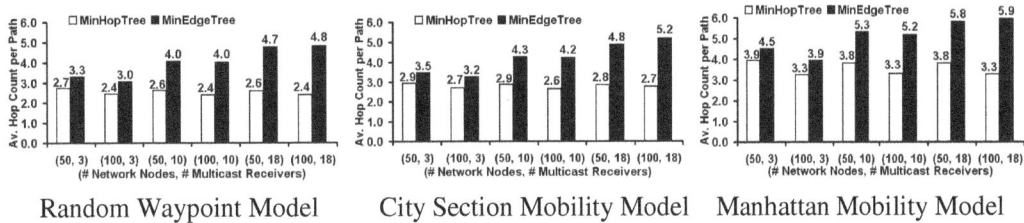

Random Waypoint Model   City Section Mobility Model   Manhattan Mobility Model

Figure 12: Hop Count per Path under Different Mobility Models (Max. Node Velocity: 5 m/s)

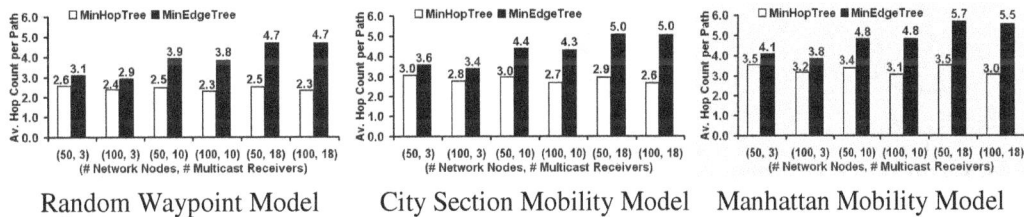

Random Waypoint Model   City Section Mobility Model   Manhattan Mobility Model

Figure 13: Hop Count per Path under Different Mobility Models (Max. Node Velocity: 25 m/s)

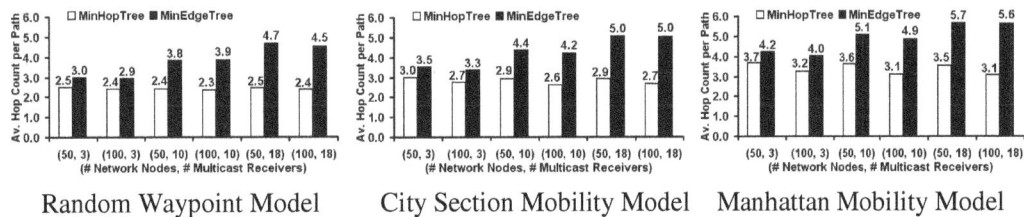

Random Waypoint Model   City Section Mobility Model   Manhattan Mobility Model

Figure 14: Hop Count per Path under Different Mobility Models (Max. Node Velocity: 50 m/s)

When compared to the average hop count per source-receiver path incurred by minimum hop trees, the average hop count per source-receiver path for minimum edge trees is 20% (for smaller number of receivers per multicast group) to 100% (for larger number of receivers per multicast group) more. Note that with increase in the network density and/or the number of receivers per multicast group, the trend of the hop counts per source-receiver path for minimum hop trees is to decrease; whereas, the trend of the hop count per source-receiver path for minimum edge trees is to increase. The hop count per source-receiver path for minimum hop trees decreases by at most 14% and 30% respectively; whereas, the hop count per source-receiver path for minimum edge trees increases by at most 47%.

5.3. Lifetime per Multicast Tree

The minimum edge multicast trees had a relatively longer lifetime compared to the minimum hop multicast trees. This could be attributed to (i) the increased number of edges (refer to Section 5.2 for more on this observation) in a minimum hop multicast tree; (ii) the physical Euclidean distance between the constituent nodes of an edge on a minimum hop path is close to the transmission range of the nodes at the time of tree formation itself. Thus, the probability of an edge failure is quite high at the time of formation of the tree; (iii) the edges of a tree are also independent from each other. All these three factors play a significant role in the relatively lower lifetime per minimum hop multicast tree. While both the minimum hop and minimum edge trees exist for a relatively longer time under the Manhattan mobility model; the lifetime of the trees was the least under the City Section model for most of the scenarios.

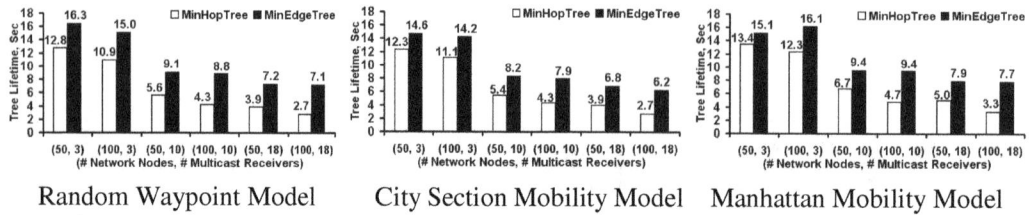

Figure 15: Lifetime per Tree under Different Mobility Models (Max. Node Velocity: 5 m/s)

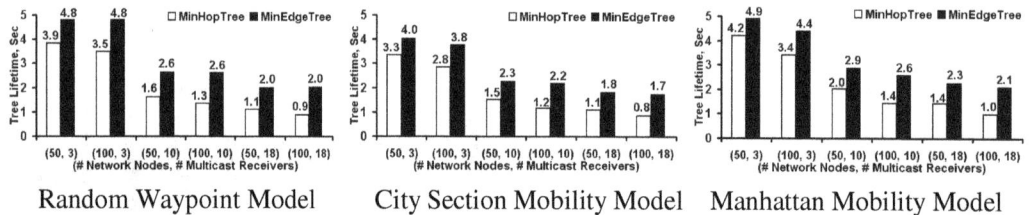

Figure 16: Lifetime per Tree under Different Mobility Models (Max. Node Velocity: 25 m/s)

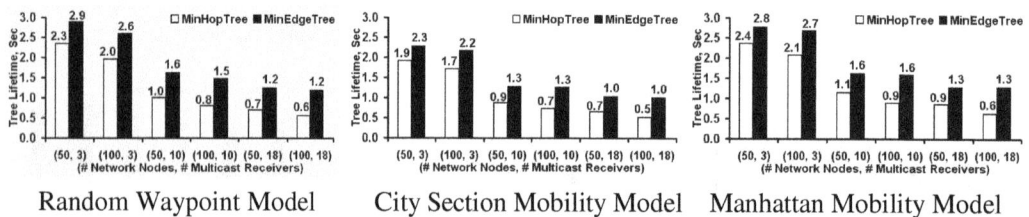

Figure 17: Lifetime per Tree under Different Mobility Models (Max. Node Velocity: 50 m/s)

For both the multicast algorithms, for a fixed network density, as the number of receivers per multicast group is increased, the lifetime per multicast tree decreases moderately at low node mobility and decreases drastically (as large as one-half to one-third of the value at smaller number of receivers per group) at moderate and high node mobility scenarios. This could be attributed to the difficulty in finding a tree that would keep the source node connected to the receivers of the multicast group for a longer time, with increase in node mobility and/or the number of receivers per multicast group. For a given number of receivers per multicast group and node mobility, the lifetime per minimum hop trees and minimum edge trees slightly decreases as we double the network density. The decrease is more predominant for minimum hop trees and this could be attributed to the relatively unstable minimum hop paths in high density networks (refer Section 5.2 for more discussion on this observation).

For a given level of node mobility, the lifetime per minimum edge tree is 23% (low density) to 38% (high density); 61% (low density) to 107% (high density) and 76% (low density) to 160% (high density) larger than the lifetime per minimum hop tree for small, moderate and larger number of receivers per multicast group respectively. For both minimum hop and minimum edge trees, for a given network density and number of receivers per group, as we increase the maximum node velocity to 25 m/s and 50 m/s, the lifetime per tree reduces by $1/3^{rd}$ to $1/6^{th}$ of their value at a maximum node velocity of 5 m/s.

6. CONCLUSIONS

We have described the algorithms that can be used to obtain benchmarks for the minimum hop count per source-receiver path and minimum number of edges per tree for multicast routing in mobile ad hoc networks. Simulations have been conducted to obtain such benchmarks for different conditions of network density, node mobility and number of receivers per multicast group under three different mobility models – the Random Waypoint model (used for MANETs) plus the City Section and Manhattan model (used for VANETs). Both the minimum edge and minimum hop based multicast trees are inherently more stable under the Manhattan model and least stable under the City Section model. The Random Waypoint model supports the minimum edge trees and minimum hop trees to have the lowest values for the number of edges and hop count per source-receiver path metrics.

For a particular mobility model, the minimum hop based multicast trees have a larger number of edges than the theoretical minimum – the minimum hop trees are unstable and their lifetime decreases with increase in the number of edges. This could be attributed to the instantaneous decision taken by the minimum hop path algorithm to select a tree without any consideration for the number of edges and their lifetime. The minimum edge trees have a relatively larger hop count per source-receiver path and the hop count per path increases with the number of receivers per multicast group. The relatively fewer edges in the minimum edge tree results in a relatively larger lifetime compared to the minimum hop trees, as each edge in these two trees are independent. The simulation results thus indicate a complex tradeoff between the hop count per source-receiver paths and number of edges per tree vis-à-vis their impact on the lifetime per tree for multicast routing.

REFERENCES

[1] C. Siva Ram Murthy and B. S. Manoj, "Routing Protocols for Ad Hoc Wireless Networks," *Ad Hoc Wireless Networks: Architectures and Protocols*, Chapter 7, pp. 299 – 364, Prentice Hall, 1st Edition, June 2004.

[2] C. K. Toh, G. Guichal and S. Bunchua, "ABAM: On-demand Associatvity-based Multicast Routing for Ad hoc Mobile Networks," *Proceedings of the 52nd IEEE VTS Fall Vehicular Technology Conference*, Vol. 3, pp. 987 – 993, September 2000.

[3] T. H. Cormen, C. E. Leiserson, R. L. Rivest and C. Stein, "Introduction to Algorithms," 2nd Edition, MIT Press/ McGraw-Hill, Sept. 2001.

[4] L. Kou, G. Markowsky and L. Berman, "A Fast Algorithm for Steiner Trees," Vol 15, pp. 141-145, *Acta Informatica*, Springer-Verlag, 1981.

[5] N. Meghanathan, "On the Stability of Paths, Steiner Trees and Connected Dominating Sets in Mobile Ad hoc Networks," *Ad Hoc Networks*, Vol. 6, No. 5, pp. 744-769, July 2008.

[6] K. Fall, K. Varadhan, "The ns Manual," The VINT Project, A Collaboration between researchers at UC Berkeley, LBL, USC/ISI and Xerox PARC.

[7] A. Farago and V. R. Syrotiuk, "MERIT: A Scalable Approach for Protocol Assessment," *Mobile Networks and Applications*, Vol. 8, No. 5, pp. 567 – 577, October 2003.

[8] C. Bettstetter, H. Hartenstein and X. Perez-Costa, "Stochastic Properties of the Random-Way Point Mobility Model," *Wireless Networks*, pp. 555 – 567, Vol. 10, No. 5, September 2004.

[9] T. Camp, J. Boleng and V. Davies, "A Survey of Mobility Models for Ad Hoc Network Research," Wireless Communication and Mobile Computing, Vol. 2, No. 5, pp. 483-502, September 2002.

[10] F. Bai, N. Sadagopan and A. Helmy, "IMPORTANT: A Framework to Systematically Analyze the Impact of Mobility on Performance of Routing Protocols for Ad hoc Networks," *Proceedings of the IEEE International Conference on Computer Communications*, pp. 825-835, March-April, 2003.

[11] A. Vasiliou and A. A. Economides, "Evaluation of Multicast Algorithms in MANETs," *Proceedings of the 3rd World Enformatika Conference – International Conference on Telecommunications and Electronic Commerce*, vol. 5, pp. 94-97, Stevens Point, Wisconsin, USA, 2005.

[12] R. Biradar, S. Manvi and M. Reddy, "Mesh Based Multicast Routing in MANETs: Stable Link Based Approach," *International Journal of Computer and Electrical Engineering*, vol. 2, no. 2, pp. 371-380, April 2010.

[13] E. Menchaca-Mendez, R. Vaishampayan, J. J. Garcia-Luna-Aceves and K. Obraczka, "DPUMA: A Highly Efficient Multicast Routing Protocol for Mobile Ad Hoc Networks," *Lecture Notes in Computer Science*, vol. 3738, pp. 178-191, 2005.

[14] E. Baburaj and V. Vasudevan, "Exploring Optimized Route Selection Strategy in Tree- and Mesh-Based Multicast Routing in MANETs," *International Journal of Computer Applications in Technology*, vol. 35, no. 2-4, pp. 174-182, 2009.

[15] N. Meghanathan and S. Vavilala, "Impact of Route Selection Metrics on the Performance of On-Demand Mesh-based Multicast Ad hoc Routing," *Computer and Information Science*, vol. 3, no. 2, pp. 3-18, May 2010.

[16] N. Meghanathan, "Multicast Extensions to the Location-Prediction Based Routing Protocol for Mobile Ad hoc Networks," *Lecture Notes of Computer Science*, vol. 5682, pp. 190-199, August 2009.

[17] S. S. Manvi and M. S. Kakkasageri, "Multicast Routing in Mobile Ad hoc Networks by using a Multiagent System," *Elsevier Information Sciences*, vol. 178, no. 6, pp. 1611-1628, March 2008.

[18] P. Kamboj and A. K. Sharma, "Scalable and Robust Location Aware Multicast Algorithm (SRLAMA) for MANET," *International Journal of Distributed and Parallel Systems*, vol. 1, no. 2, pp. 10-24, November 2010.

8

A Handoff-based And Limited Flooding (HALF) Routing Protocol in Delay Tolerant Network (DTN)

Anika Aziz[1] and Shigeki Yamada[2]

[1]Applied Physics, Electronics & Communication Eng, University of Dhaka, Dhaka, Bangladesh
anika.aziz@gmail.com
[2] National Institute of Informatics, 2-1-2 Hitotsubashi, Chiyoda-ku, Tokyo 101-8430, Japan
shigeki@nii.ac.jp

ABSTRACT

In a Delay Tolerant Network (DTN), routing protocols are developed to manage the disconnected mobile nodes. We propose a routing protocol named HALF (Handoff-based And Limited Flooding) in DTN that can work in both infra-structured and infra-structure less networking environment and hence it can improve the performance of the network significantly. In this paper, it is shown that HALF gives satisfactory delivery ratio and latency under almost all conditions and different network scenarios when compared to the other existing DTN routing protocols. As the traffic intensity of the network grows from low (.2) to high (.75) values, HALF shows about 5% decrease in the delivery ratio compare to much larger values showed by the other routing protocols and on the average takes same time to deliver all the messages to their destinations. As the radio range is increased over the range from 10m Bluetooth range to 250m WLAN range, due to the increased connectivity, the delivery ratio and the latency are increased by 4 times and decreased by 5 times respectively.

Keywords
DTN, handoff, flooding, DTN routing protocols, infra-structured networking, infra-structure less networking, HALF.

1. INTRODUCTION

Delay and Disruption Tolerant Networking (DTN) [1] equipped with advance features as custody transfer and hop-by-hop routing capabilities give a full potential of flexibility, adaptability and simplicity for wider range of different characteristics of network. The custody transfer capability allows messages or Bundles to be buffered in DTN nodes until Bundles are forwarded to the next hop DTN node and found to be unnecessary. The hop-by-hop routing capability enables routing decisions to be made dynamically during each hop [1]. HALF (Handoff-based And Limited Flooding) is an integrated routing scheme that combines an infrastructure-oriented DTN routing scheme with a flooding technique that works well for an infrastructure-less environment. The infrastructure-based routing scheme of HALF is a Handoff–based routing protocol that makes the best use of general handoff mechanisms intended for the IP network. In HALF, this handoff mechanism is implemented using the DTN features like hop-by-hop routing and custody transfer. For an infrastructure-less environment, HALF applies a flooding technique similar to Spray and Wait (SW) [2] protocol to spread message in the network but in a more controlled way.

Other existing DTN routing protocols like Epidemic [3], PRoPHET [4] and SW [2] are basically flooding based and mobility dependent routing protocols that simply utilize the local knowledge given by adjacent nodes but do not utilize the global connectivity knowledge on fixed network topology. These protocols are suitable for a network consisting mobile nodes only. On the other hand, HALF is suitable to be used in any type of networking environment whether it has only mobile nodes or both mobile and fixed nodes. In this paper, at first we explained our proposed HALF routing protocol and then the performance characteristics of HALF with the existing DTN routing protocols are intensively compared under a wide variety of network environments and conditions. The simulation results indicate that in most of the cases, HALF achieves higher delivery ratio and lower end-to-end latency under broad network environment in comparison to others.

The rest of the paper is organized as follows: Section 2 presents related work. Section 3 gives the basic mechanism and Protocol operation of HALF. Section 4 presents the Performance evaluation and analysis. Section 5 concludes the paper.

2. RELATED WORK

2.1. Related work on handoff technologies in TCP/IP protocol

Different methods were devised to overcome the problems associated with the TCP to handle mobility in the wireless environment [5], [6], [7] and [8] and to handle the handoff situations efficiently [9], [10], [11]. Protocols like Mobile IP[12] suffers from scalability problem and Cellular IP [13], [14] accompanies additional network load induced by forwarding packets on multiple paths and sometimes may cause packet loss due to the transient packet transfers to the old route. Also IP mobility-based techniques need explicit buffering instruction to the routers during handover to buffer packets at the router. Protocols like HALF in DTN does not have the end-to-end session management or connection state transfer problem during handoff and have lower handoff latency and overall latency than Mobile IP protocol. This is because handoff process in HALF implements handoff of the messages with minimum number of control message exchanges between the fixed nodes without transferring session state information that is necessary to keep an end-to-end TCP session. Furthermore, the custody transfer mechanism of DTN does not require any extra overhead of explicitly instructing to buffer packets during the handover process. Also the Custody Acceptance signalling [15] can control the burst packet transfers and so HALF does not suffer from the multiple consecutive packet loss problems which is unlike in TCP/IP-based forwarding and buffering scheme.

2.2. Related work on DTN routing protocols

In this section we present a brief overview of DTN routing techniques relevant to our proposed protocol. Existing protocols in DTN were designed to handle the challenging and opportunistic situations of sparsely connected Mobile nodes in a network. Epidemic routing protocol is solely based on the information exchanges between two encountering mobile nodes and thus distributing the messages throughout the network to reach the destination. The PRoPHET was devised to be more selective by being probabilistic while forwarding to the next node. The Spray and Wait (SW) protocol adds limited copy flooding feature to the mobile nodes while routing to the destination. These flooding based routing protocols do not make use of the global knowledge and hence suffers from reduced delivery ratio and large latencies. MaxProp prioritizes the scheduling of packets for the transmission and take the resource limitation into account. HALF assumes simple FIFO for scheduling the packets. Another DTN routing protocol, RAPID, deals with the problem of routing in DTN as a resource allocation problem and tries to solve it by calculating a routing metric per packet on the basis of available resources and then replicate the packet according to that. HALF does not have any replication method in its operation neither does it involves calculating the routing metric in the basis of resources available.

3. THE HANDOFF-BASED AND LIMITED FLOODING (HALF)

3.1. The basic mechanism

HALF makes use of the general handoff mechanisms intended for the IP network but uses DTN features like hop-by-hop routing and custody transfer. The knowledge of the location of the mobile node is utilized. The route update information during handoff and Back Propagation and caching of this location information over the experienced route helps to route the Bundle to the destination quickly and deterministically. A limited flooding technique is integrated to this mechanism, resulting in a much improved routing protocol that fits to a wide range of scenarios. To implement these concepts in our routing protocol, we extend the Bundle Protocol's message format given by the IRTF's Delay Tolerant Network Research Group (DTNRG) as BP specification [15]. The routing functions and handoff mechanism are included in the message format so that the routing and handoff can be implemented in a single unified layer. The unified BP layer stands between the Link layer and Application layer. It provides the reliable transfer and dynamic routing through hop-by-hop dynamic next hop selection and also efficient buffering mechanism during handoff through custody transfer.

3.2. The protocol operation

When a mobile node moves out from the coverage of a fixed router to another one, Handoff process takes place. Normally, every DTN router maintains the connectivity information with adjacent DTN routers in the Proxy List (PL). When a mobile node moves to a new location, it registers its location (the name of the DTN router it belongs to) with the new DTN router and this location information is propagated back to and cached in every DTN router over the experienced route to update the PL at each router. During this process, each of the DTN routers also maintains a Back List (BL) to keep the information of the old router of a mobile node to track the experienced route of that mobile node. So a router receiving a Bundle to be delivered to a mobile node for which it has cache route update information, can make use of the information to route the Bundle to the destination quickly and deterministically.

This improves the delivery ratio of the network, preserving the low overall latency. By increasing the cache time at each of the router it is also possible to increase the delivery ratio. Figure 1 shows the back propagation and caching with the PL and BL. If a node does not have information about the destination then it starts flooding but if any of the branch node has information in its PL about the destination then it need not to do the flooding. Instead, that node will forward the Bundle to the proxy found in the PL. That is why we termed it as Limited Flooding (LF).

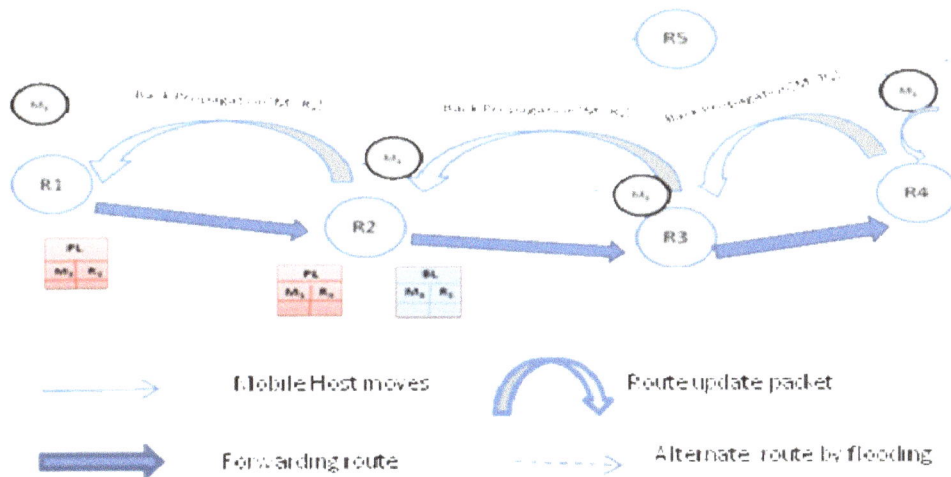

Figure1. Back propagation of routing updates using BL and caching at the Routers using PL

We give two examples in Table 1, from our simulation traces to show how the messages are transmitted through a network by selecting Proxy (PX) or Limited Flooding (LF) method depending upon the ongoing situation.

Table 1: Messages in an infrastructure-based and infrastructure-less network

Message	Host	Transmission type	Time	Remarks
Message **M5** for Infrastructure-based network	W49 (Walker)	CT (Creation Time)	22	In case of infrastructure-based network most of the messages find proxies on their way to the destination with few steps of flooding.
	P5 (Pedestrian)	LF	25.5	
	Fixed nodes :@116	LF	29.1	
	@115	LF	32.6	
	@107	PX	36.1	
	@106	PX	44.2	
	@103	PX	53.2	
	@80	PX	56.8	
	@79	DR (Direct Transmission)	63.3	
Message **M25** for Infrastructure-less network	@125	CT	82	For an infrastructure-less environment most of the transmission from one hop to next hop is by Limited Flooding method.
	@66	LF	85.9	
	@95	LF	89.9	
	P1	LF	93.7	
	t 63 (tram)	LF	97.8	
	@101	DR	185.8	

From Table 1, Message M5 was created at W49 at 22nd instant of time. It was delivered to P5 by SW flooding method at 25.5th sec. M5 was delivered to Fixed routers @116 and @115 by similar method. @115 found a proxy (PX) to the destination that is @107 and so M5 is delivered to @107 from @115 by proxy (PX) method. Finally fixed router @80 could send M5 by Direct Transmission to the final destination, @79. The second example of M44 has a similar explanation but there is no transmission by PX method.

The forwarding mechanism is such that a router always looks for a direct connection while forwarding a Bundle. If it is not found then the router consults the PL and lastly it goes for the flooding technique as shown in Fig. 2.

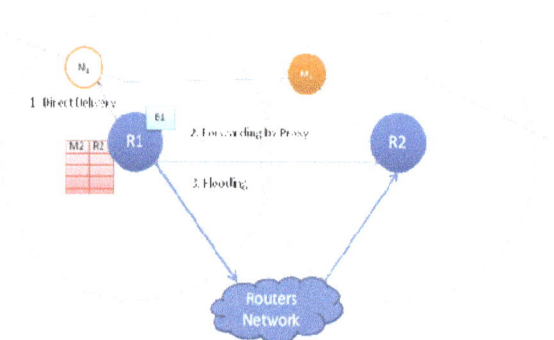

Figure 2. Basic forwarding mechanism in HALF

4. THE HANDOFF-BASED AND LIMITED FLOODING (HALF)

4.1. Network simulation model

Our network model, shown in Fig. 3, consists of Fixed and Mobile nodes. The mobile nodes such as Walkers, Pedestrians, Cars and Trams are plying along different routes in the Helsinki City map as featured in ONE simulator following a Map based Movement Model [16]. Real world aspects are added to the synthetic mobility models by adding real-world street maps, different classes of mobile nodes, realistic connectivity etc. [17]. HALF Routing protocol extends the Active Router module used in the ONE simulator. Fields and methods have been created to implement the Handoff mechanism which was not included in any DTN routing algorithm in ONE before. Special Reports have been generated by extending the Report module.

Figure 3. A General network model with fixed and mobile routers

Fixed nodes are connected with each other through communication link of 250 kBps (kilo Bytes per second). To test the wider applicability of our routing protocol we have simulated it and made comparison with other existing routing protocols in different types of network models: Mostly Fixed, Mixed, Mostly Mobile and All-Mobile where the number of fixed nodes with respect to the number of the mobile nodes are kept higher, equal, smaller and nil respectively. Both types of nodes are varied for 35, 50 and 65 numbers among 100 nodes in total within (4500 x 3400) m simulation area. The TTL of the message is 40mins (for discarding messages). The simulation time was for 12 hours. For every simulation case, we have chosen five runs using different random seeds and report the average value. As a performance metrics for evaluation, we have used delivery ratio and average latency.

It is very important to study how a protocol can handle different types of networks, different traffic load conditions of the network and different message sizes. The buffer size at different nodes of a network influences the perfromances of a routing protocol for a particular traffic and message size conditions. It is also required to see how the different radio ranges affect the number of delivered bundle and time to be delivered to their destinations. While the existing routing protocols were developed for sparse mobile environment only, in this paper we would like to study the perfromance of the routing protocols where the number of fixed nodes are kept constant but mobile nodes are varied in a network. How a random and a more realistic Mobility model along with the variation in the different mobility speed contributes to the performance of HALF and other protocols are also worthwhile to study as these also give a insight to how much message size the routing protocol can handle under different Mobility conditions.

Table 2 shows the different scenarios that we have simulated.

Table 2. Different scenarios simulated for our model

Scenarios	Parameters	Details
Mostly Fixed Routers Fixed=Mobile Routers Mostly Mobile Routers All Mobile Routers	Traffic Load of the network	Message generation interval of [1, 29], [1, 11] and [1, 7] corresponds to Traffic intensity (ρ) value of low (0.2), medium (0.5) and high (0.75) respectively.
	Message sizes	Varied as [100KB - 2MB], [500KB - 4MB], [500KB - 8MB] and [1MB - 100MB] with [1, 29] interval.
	Buffer sizes	Pedestrians, Walkers and Cars have 5Mbytes, Fixed nodes have 20Mbytes and the Trams have 50Mbytes each. These values are increased to 10M, 100M and 100M respectively.
All Mobile Routers	Radio ranges	10m~250m
	Mobility model	RWP, SPMBM
	Mobility speed	Varying the speed of the different mobile carriers in the simulation model

4.2. Simulation Results

4.2.1. Performance at different traffic intensity

It is shown in Fig. 6 and Fig. 7 that increase in traffic intensity decreases the Delivery ratio and increases the latency for both 10m and 100m radio range. Because the nodes cannot deliver the increased traffic due to overburden causes, delivery ratio decreases. On the other hand, as the traffic intensity increases the average time to reach the destination for the messages increases due to increased waiting time but this increase in the latency is a gradual increase rather than having a sharp profile. This behavior is because of the contribution from the encountering delay between two nodes out of the total delay. As we go from the Mostly fixed to the All mobile network, the delivery ratio decreases because of less contribution from the interconnected fixed nodes and the latency increases as the bundles can reach to their destination only by the movement of the mobile nodes. It is noteworthy that HALF gives higher delivery ratio and lower latency than all other protocols under almost all above mentioned conditions and scenarios. It is found that HALF is suitable for 100m wireless range than 10 m wireless range due to wider coverage of moving nodes.

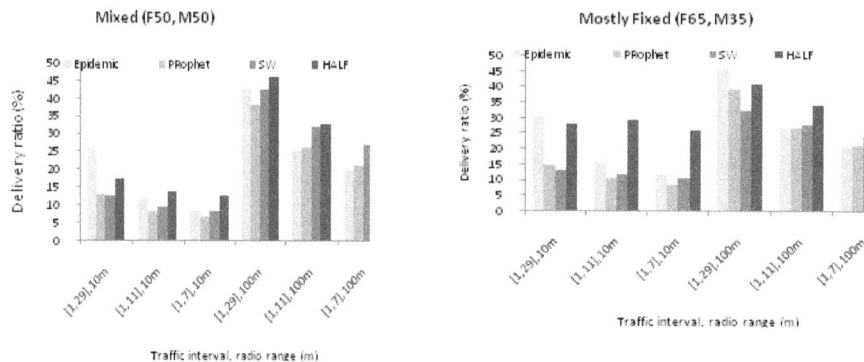

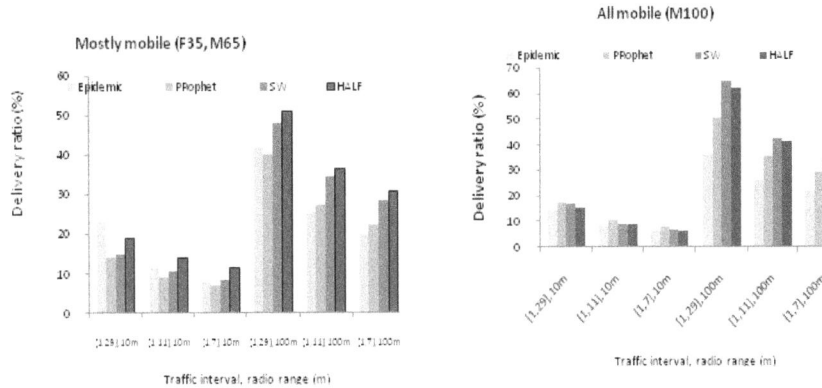

Figure 6. Delivery ratio with different number of Fixed and Mobile nodes

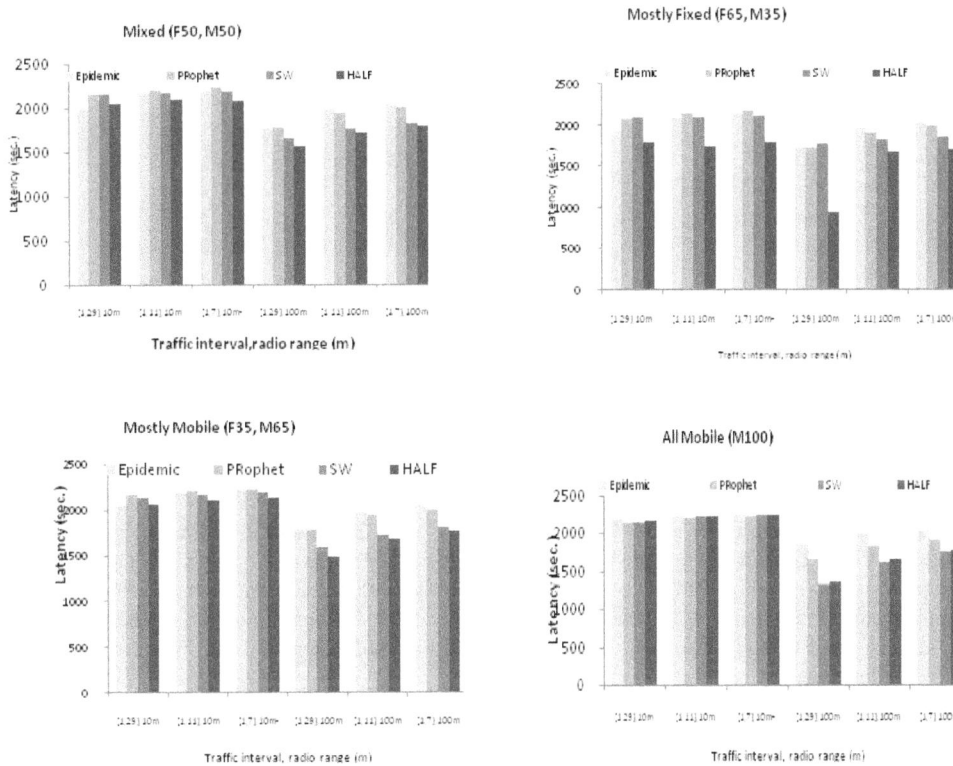

Figure 7. Latency with different number of Fixed and Mobile nodes

4.2.2. Different Message sizes

The Pedestrians, Walkers and Cars have the Buffer size of 5Mbytes, the Fixed nodes having buffer of 20Mbytes and the Trams are having 50Mbytes each. Because of the opportunistic contacts, larger messages cannot be always successfully delivered. So, the Delivery ratio decreases as the Message size increases for all the protocols, for both types of scenario as shown in Fig. 8. Thanks to the support for the fixed infrastructure, the delivered bundles take less overall time for Mostly Fixed environment but in an All Mobile scenario the latency is higher. With the increase of the Message size the latency decreases as less number of bundles takes less time to be delivered to their destination. Interestingly, HALF gives much better performance than the other protocols in a Mostly fixed environment but in an All mobile scenario its performance is a bit lower than other protocols.

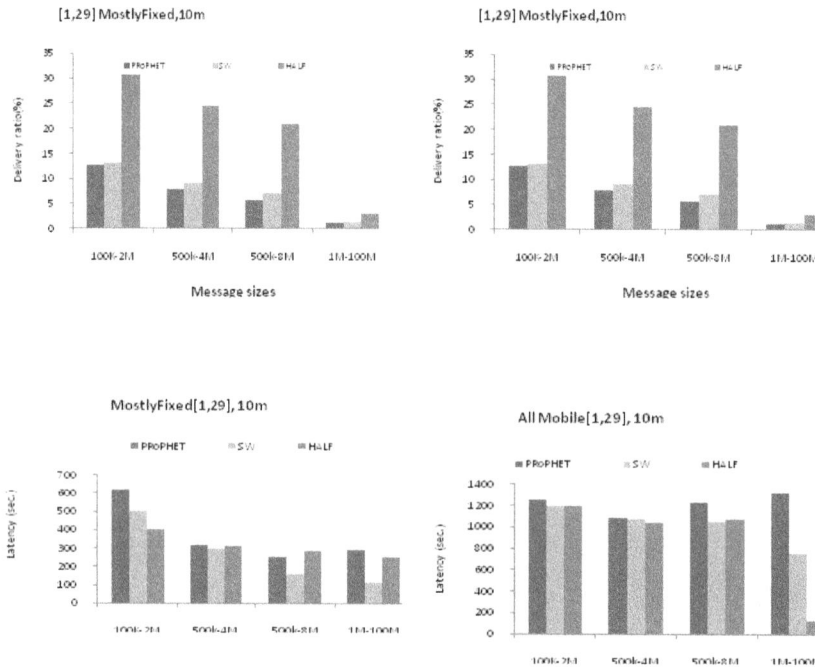

Figure 8. Performance for different Message sizes

4.2.3. Different Buffer sizes at the nodes

The buffers at the nodes (as mentioned previously) are increased to 10M, 100M and 100M respectively, for a particular Message size of [500k, 1M] and for low traffic [1, 29] and high traffic [1, 7] conditions under the Mostly Fixed scenario. As shown in Fig. 9, the increased Buffer size at each node causes the delivery ratio to be increased by 50% because now more bundles can be buffered at the nodes to wait for the next opportunity to be delivered instead of getting dropped. At the same time, the bundles now take longer to get delivered to their destination because of increased buffering time which leads to reduced value of overall latency.

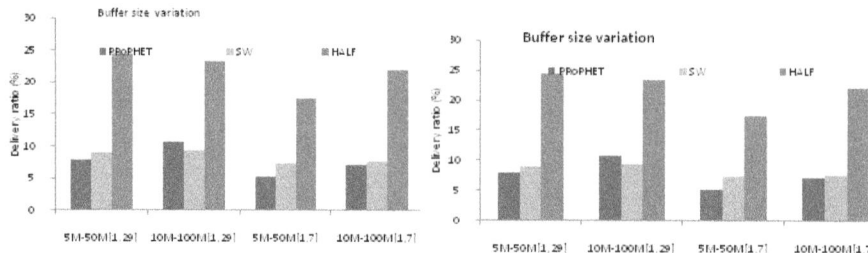

Figure 9. Performance for different Buffer sizes at the nodes

4.2.4. Different Radio ranges

We showed in Fig. 10 how HALF and other protocol behaves for different radio ranges starting from Bluetooth (10m) range to Wireless LAN range (100m) and even larger ranges like 200/250m, considering the futuristic probability of using higher wireless range devices for communication. As the communication range increases the connectivity among the nodes increases. As a result of this, the delivery ratio and the latency at 250m wireless range is increased by 4 times and decreased by 5 times respectively than their value at 10m wireless range. When the buffer size is increased to [10/100/100] M at different nodes, for 10m, 100m and 250m ranges, it is found in Fig. 10 that more bundles are delivered because of the increased Buffers but are taking longer time to reach their destination

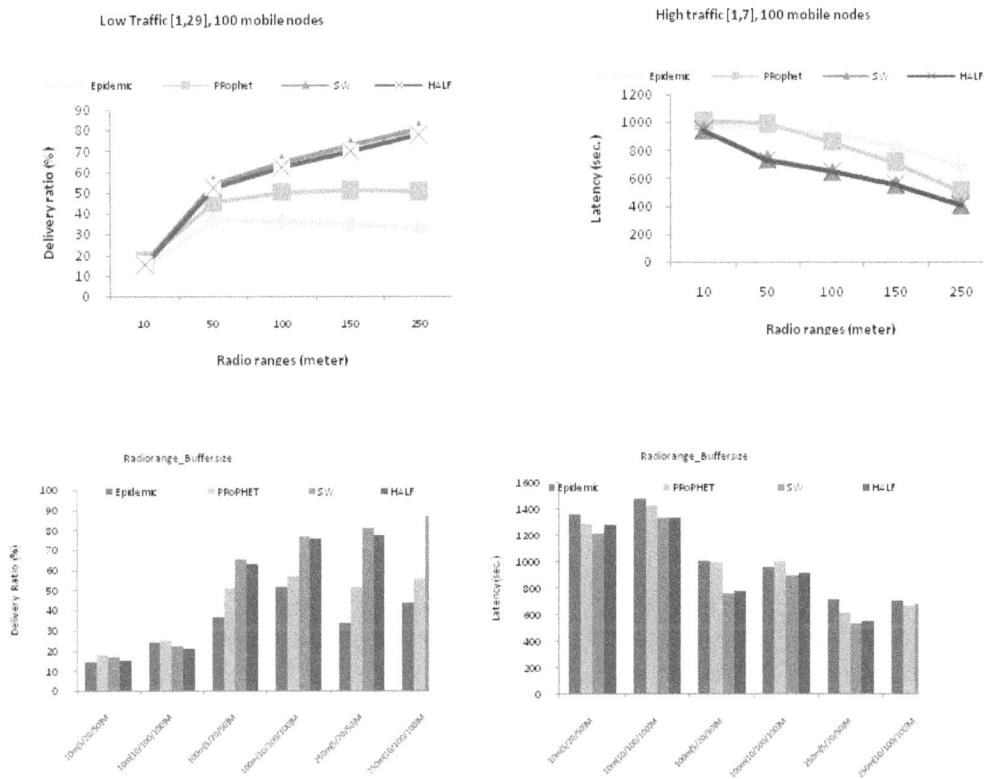

Figure 10. Performance at different Radio ranges and with different
Buffer sizes at the nodes

4.2.5. Different Mobility model

To study how the Mobility models can affect the different protocol performances for the All Mobile network environment, we try with one random model like Random Way Point (RWP) and another more realistic model like Shortest Path Map Based Mobility Model (SPMBM). As shown in Fig. 11, with SPMBM the delivery ratio is higher than with RWP. The latency is best with HALF compared to other protocols. We also observed the performances by varying the number of different type of Mobile nodes. The number of cars influences the delivery performances very much because of the increased contact frequency. The number of trams has less influence on this as we found that with no trams but 40 cars the delivery ratio is better than with no Cars (but trams and others). In summary, SPMBM mobility model with high speed vehicle improves the performance of the protocol.

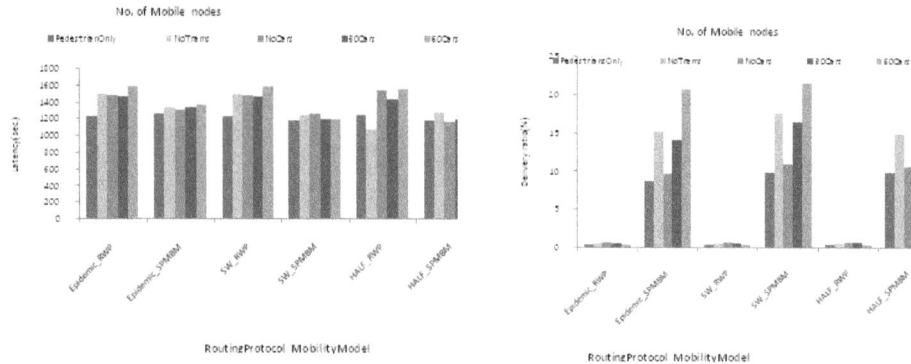

Figure 11. Performance for different Mobility models

4.2.6. Different Mobility speed

Further sub-sectioning, if required, is indicated

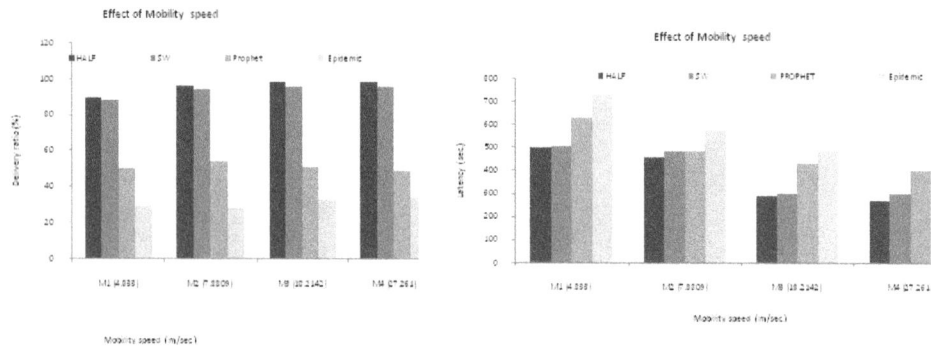

Figure 12. Effect of different Mobility speed on the

We calculated different Mobility speed M1, M2, M3 and M4 using the following relation :

$\sum$(minimum speed, maximum speed) in m/s of each type of mobile node groups × no. of hosts in that group/total no. of nodes; the range of the speed for each mobile node types were varied to calculate the values of different M.

Fig. 12 presents that as the mobility increases, the delivery ratio increases and the latency decreases. The latency of HALF decreases to about 60% of its value at M4 than at M1 because due to the moblity of the nodes more Bundles can reach their destination faster than before.

4.2.7. Different route-cache time variation

By varying the Route-Cache time of the update message at each of the routers from 5 sec to 3000 sec value range we get increased delivery ratio and decreased latency value. Because of the increased cache time, more Bundles can get their way to the destination using the route information and Bundles from the far away routers are now contributing in the latency values.

5. CONCLUSION

Our main achievement is that HALF gives satisfactory performances over a wide range of network types. Considering the latency, HALF is better than all other existing routing protocols for low to high traffic intensity conditions in the network. This protocol could also achieve higher delivery ratio than others except at low traffic intensity. HALF can work over the radio range from 10m to 250m with improved performance as the range increases. We also showed that HALF is suitable for small message sizes. The SPMBM mobility model with high speed vehicle improves the performance of the protocol and with the increase of the Mobility speed the delivery ratio increases and delay decreases to a substantial amount.

REFERENCES

[1] K. Fall, "A Delay-Tolerant Network Architecture for Challenged Internets" In ACM SIGCOMM, Aug. 2003J. Clerk Maxwell, A Treatise on Electricity and Magnetism, 3rd ed., vol. 2. Oxford: Clarendon, 1892, pp.68–73.

[2] T. Spyropoulos, K. Psounis, C. S. Raghavendra," Spray and Wait: An efficient routing scheme for intermittently connected mobile networks", SIGCOMM`05 Workshops, August 22-26, 2005

[3] A. Vahdat and d. Becker, "Epidemic routing for partially connected ad hoc networks", Technical Report CS-200006, Duke university, April,2000

[4] Lindgren, A Doria and O. Schelen, "Probabilistic routing in intermittently connected networks", SIGMOBILE Mobile Computing and Communications Review, 7(3):19-20, 2003

[5] P Manzoni, D Ghosal, G Serazzi, "Impact of mobility on TCP/IP: an integrated performance study", IEEE Journal on Selected Areas in Communications, 1995

[6] R Yavatkar, N Bhagawat, "Improving end-to-end performance of TCP over Mobile Internetworks", Mobile Computing Systems and Applications,1994

[7] H Balakrishnan, S Seshan, RH Katz, "Improving reliable transport and handoff performance in cellular wireless networks", Wireless Networks, 1995

[8] R Caceres, L Iftode, "The effects of mobility on Reliable Transport Protocols", IEEE Journal of selected areas in communication.1994

[9] Ajay V. Bakre and B.R.Badrinath, "Implementation and Performance Evaluation of Indirect TCP", IEEE transaction on computer,vol.46,no.3. March1997

[10] Balakrishnan, H., et al. Improving TCP/IP Performance over Wireless Networks. In Intl. Conf. on Mobile Computing and Networking. 1995. Berkeley,CA.

[11] K. Brown and S. Singh, M-TCP: TCP for mobile cellular networks, Computer Communication Review 27(5) (1997) 19–43.

[12] Jochen Schiller, "Mobile communication", 2nd edition, Addison-Wesley, 2003

[13] AT Campbell, J Gomez-Castellanos," IP micro mobility protocols",SIGMOBILE Mobile Computing and Communications Review,2000

[14] Andrew T. Campbell, Javier Gomez, Sanghyo Kim, Andras G. Valko, and Chien-Yin Wan, Design, "Implementation, and Evaluation of Cellular IP", IEEE Personal Communications, August, 2000

[15] K. Scott and S Burleigh, "Bundle Protocol Specification", November, 2007K.

[16] Ari Keränen, Jörg Ott and Teemu Kärkkäinen "The ONE Simulator for DTN Protocol Evaluation", SIMUTools'09: 2nd International Conference on Simulation Tools and Techniques. Rome, March 2009

[17] Ari Keränen and Jörg Ott, Increasing Reality for DTN Protocol Simulations, Tech. rep., Helsinki University of Technology, Networking Laboratory, July 2007, pages 1-9 .

9

A CROSS-LAYER QOS AWARE NODE DISJOINT MULTIPATH ROUTING ALGORITHM FOR MOBILE AD HOC NETWORKS

Mahadev A. Gawas[1], Lucy J.Gudino[2], K.R. Anupama[3], Joseph Rodrigues[4]

[1,2] Department of Computer Science BITS PILANI K.K. Birla Goa campus.

[3]Department of EEE/EI BITS PILANI K.K. Birla Goa campus.
[4]Department of Electronics ATEC Verna goa.

ABSTRACT

Future mobile Ad hoc networks (MANETs) are expected to be based on all-IP architecture and be capable of carrying multitudinous real-time multimedia applications such as voice, video and data. It is very necessary for MANETs to have a reliable and efficient routing and quality of service (QoS) mechanism to support diverse applications which have variances and stringent requirements for delay, jitter, bandwidth, packet loss. Routing protocols such as AODV, AOMDV, DSR and OLSR use shortest path with the minimum hop count as the main metric for path selection, hence are not suitable for delay sensitive real time applications. To support such applications delay constrained routing protocols are employed. These Protocols makes path selection between source and destination based on the delay over the discovered links during routing discovery and routing table calculations. We propose a variation of a node-disjoint Multipath QoS Routing protocol called Cross Layer Delay aware Node Disjoint Multipath AODV (CLDM-AODV) based on delay constraint. It employs cross-layer communications between MAC and routing layers to achieve link and channel-awareness. It regularly updates the path status in terms of lowest delay incurred at each intermediate node. The performance of the proposed protocol is compared with single path AODV and NDMR protocols. Proposed CLDM-AODV is superior in terms of better packet delivery and reduced overhead between intermediate nodes.

KEYWORDS

AODV, Cross Layer, MANET, MAC, NS2, QoS.

1. INTRODUCTION

MANETs are self-organizing, rapidly deployable wireless network that require no fixed infrastructure. It is composed of wireless mobile nodes that can be deployed anywhere, and can dynamically establish communications using limited network management. Real time applications have been most popular among the applications run by ad hoc networks. It strictly adheres to the QoS requirements such as overall throughput, end-to-end delay and power level. Traditionally multihop wireless network protocol design is largely based on a layered approach. Here each layer in the protocol stack is designed and operated independently with interfaces between layers that are rather static. This paradigm has greatly simplified network design and led to the robust scalable protocols on the internet. However, the rigidity of this paradigm results in poor performance for multihop wireless networks in general, especially when the application has high bandwidth requirements and/or stringent delay constraints [1]-[4].

1.1 RELATED WORK

To meet these QoS requirements, recent study on multihop networks has demonstrated that cross-layer design which can significantly improve the system performance [5]-[6]. To guarantee QoS in MANETs for delay sensitive applications two factors are considered. Firstly, route selection criterion must be QoS-aware i.e., it must consider the link quality before using the link to transmit. Secondly, the instantaneous response to the dynamics of MANET topology changes must be considered so that the route changes are seamless to the end user over the life time of a session. Generally, a QoS model defines the methodology and architecture by which certain types of services can be provided in the network. Protocols such as routing, resource reservation signaling and MAC must cooperate to achieve the goals set by the QoS model. QoS routing is one of the most essential parts of the QoS architecture [7]–[9]. Multipath approach has many advantages such as load balancing, QoS assurance and fault tolerance [10]- [12]. Several multipath routing protocols have been proposed so far in the literature. One of the earliest multipath routing protocols is Ad hoc On demand Multipath Distance Vector (AOMDV) [13]. AOMDV is a variant of Ad Hoc On Demand Distance Vector (AODV) [14] which establishes loop-free and link-disjoint paths based on the minimum hop count. QoS AODV (QS-AODV) in [15] extended the basic AODV routing protocol to provide QoS support in MANETs. It uses hop count as criterion for choosing the route with an assumption that NODE_TRAVERSAL_TIME (NTT) is constant. Stephane Lohier et al.[16] proposed reactive QoS routing protocol that also deals with delay and bandwidth requirements. In his proposal, QoS routes are traced by node to node and NTT is an estimate of the average one-hop traversal time, which includes queue, transmission, propagation, and other delays.

Cross-layered multipath AODV (CM-AODV)[17], selects multiple routes on demand, based on the signal-to-interference plus noise ratio (SINR) measured at the physical layer. Load Balancing AODV (LBAODV)[18] is a new multipath routing protocol that uses all discovered path simultaneously for transmitting data. By using this approach data packets are balanced over discovered paths and energy consumption is distributed across many nodes throughout the network.

Xuefei Li et al. [19] proposed Node-Disjoint Multipath Routing protocol (NDMR) by modifying and extending AODV to enable the path accumulation feature of DSR in route request packets. Multiple paths between source and destination nodes are discovered with low broadcast redundancy and minimal routing latency. A delay aware protocol proposed in Boshoff et al. [20], uses end-to-end delay, instead of hop count, as metric for route selection. Upon route failure, the route table which contains multiple paths, along with the end-to-end delay is first searched for an alternative route to the destination before a new route discovery process is initiated. Even though it reduces both routing overhead and end-to-end packet delay, the route delay information might not always be upto date. Perumal Sambasivam et al. [21] modified the AODV protocol's route discovery mechanism by incorporating multiple node-disjoint paths for a particular source node along with mobility prediction.

Thus, it is found that most approaches to multipath routing protocols consider the end-to-end delay. They do not emphasize on considering the processing delay incurred at each node which may indicate the congestion or link quality along the path which is node disjoint. They also do not have a mechanism to handle expiry of stale cached routes in the route table before making their selection. Hence we propose a new algorithm CLDM-AODV with cross-layer communications between MAC and Routing layers to achieve link and channel-awareness. In section II, we describe the proposed algorithm. We present simulation results in section III followed by conclusion.

2. PROPOSED CLDM-AODV ROUTING ALGORITHM

The proposed algorithm considers only node disjoint routes which satisfy the end-to-end delay specified in the route request. For calculating end-to-end delay, the algorithm estimates inter-node packet processing delay at each node. Source node makes a selection of primary path out of available multiple QoS enable paths. The proposed algorithm includes calculation of inter-node packet processing delay at each mobile node, initiation of route discovery and route reply processes.

2.1. END-TO-END DELAY

In general, total latency or delay experienced by a packet to traverse the network from source to destination may include routing delay, propagation delay and processing or node delay. Routing delay is the time required to find the path from source to destination. Propagation delay is related to propagating bits through wireless media. Processing delay involves the protocol processing time at node x for link between node x and node y. The end-to-end delay of a path is the sum of all the above delays incurred at each link along the path [17]. For MANETs, propagation delays are negligibly small and almost same for each hop along the path. The major factors involved in computation of processing delay are the queuing delay and delay incurred at the MAC layer processing.

In the proposed method, we have named processing delay as Packet Processing Delay (PPD) which includes queuing delay and delay incurred at the MAC contention. IEEE 802.11 MAC with the distributed coordination function (DCF) is used as MAC protocol and the access method is Carrier Sense Multiple Access/Collision Avoidance (CSMA/CA) with acknowledgments. To transmit packets, nodes make use of request-to-send (RTS), clear-to-send (CTS), data and acknowledgement (ACK) packets. The amount of time between the receipt of one packet and the transmission of the next is called a short inter frame space (SIFS). Average queuing Delay at the node i is $\overline{D_i}$ is given by equation [22],

$$\overline{D_i} = \alpha \overline{D_{j-1}} + (1-\alpha)\overline{D_j} \tag{1}$$

where,

$$\alpha = \frac{(queue_{size} - queue_{length})}{queue_{size}} \tag{2}$$

$queue_{size}$ is the current size of the queue at node i, $queue_{length}$ is the length of the queue at node i and j is the current period.
The channel occupation due to MAC contention is given by,

$$T_{mac} = T_{RTS} + T_{CTS} + 3 * T_{SIFS} + T_{acc} \tag{3}$$

T_{RTS} and T_{CTS} are the time periods on RTS and CTS respectively and T_{SIFS} is the SIFS period. T_{acc} is the time for channel contention. The Packet Processing Delay (PPD) is given by:

$$PPD = \overline{D_i} + T_{mac} \tag{4}$$

2.2. ROUTE DISCOVERY

Generally in reactive protocols[1], when a source node 'S' has to communicate with destination node 'D', it initiates path discovery by broadcasting a route request packet RREQ to its neighbours. The <source-address, broadcast-id> pair is used to identify the RREQ uniquely. In the proposed system, during initial route discovery phase, more than one node disjoint path between the source and destination is determined and optimal path which satisfies QoS delay requirement is chosen for the data transmission. When this primary path breaks due to nodes mobility or path fails to satisfy QoS requirement, then one of the alternate path is chosen as the next primary path and data transmission can continue without initiating another route discovery thus reducing the overhead of additional route discovery. In the proposed algorithm, the RREQ packet is modified to contain the address of the source through which it is forwarded. The packet header contains additional field for PPD and Thresh_Delay. PPD is initialized to zero and subsequently updated at each intermediate node as per Eq.(4). Thresh_Delay is set to the maximum allowable time delay for any path from source to destination. Since RREQ is flooded network-wide, a node may receive multiple copies of the same RREQ. After receiving the first RREQ, an intermediate node can receive and collect subsequent RREQ copies for the predetermined time duration, RREQ_WAIT_TIME, which is assumed as 20ms. The intermediate node also maintains RREQcounter to limit the number of RREQ that it can receive. In our proposed system, we initialize RREQcounter to three which is as shown in Figure 1. On receiving up to three RREQs, the route with minimum PPD selected which ensures the path with highest quality. Before forwarding the RREQ, intermediate node computes its PPD and compares it with Thresh_Delay. If the difference between the Thresh_Delay and current value

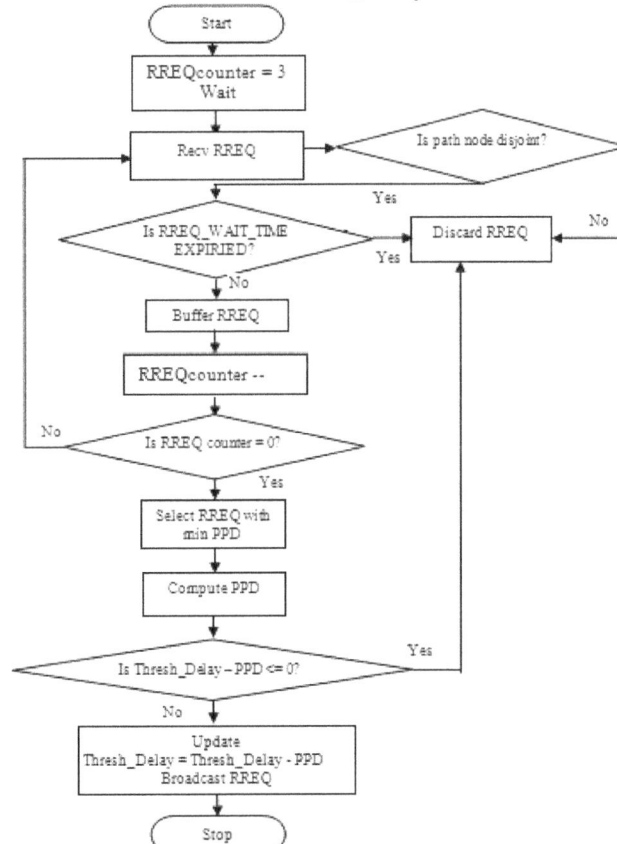

Fig. 1: RREQ Flowchart of proposed CLDM-AODV

of its PPD is zero or negative, it drops the RREQ packet avoiding unnecessary flooding   into the   network. If it satisfies, node broadcasts the packet by updating Thresh_Delay value less by currently computed PPD value of the node. Since every intermediate node forwards only one RREQ towards the destination, each RREQ arriving at the destination has traveled along a unique path from source to destination. Figure 2 shows an example of the delay based route discovery. Source node S initiates route request by updating the Thresh_Delay in RREQ Packet

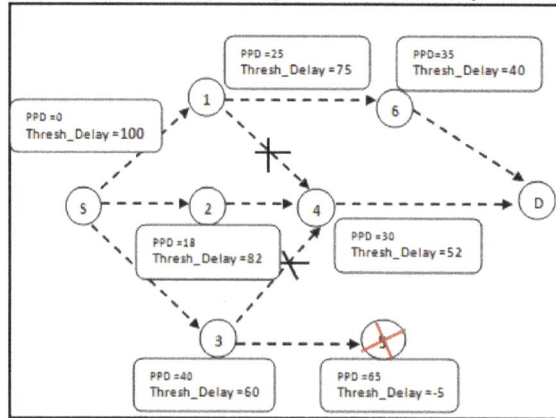

Fig. 2: Route Discovery of proposed CLDM-AODV

to acceptable delay say 100ms and PPD to zero. On receiving RREQ, node 1, 2 and 3 computes their PPD and updates Thresh_Delay in respective RREQ packet. Node 4 receives three RREQs, from node 1, node 2 and node 3 respectively. PPD values of these RREQs are compared and minimum PPD path from node 2 is chosen. Node 4 broadcast the RREQ, since it's computed PPD value satisfies the QoS constraint i.e. the difference between Thresh_Delay and PPD of node 4 is greater than zero. On the other hand, at node 5, RREQ packet gets dropped as difference between Thresh_Delay and PPD at node 5 do not satisfy the QoS criteria. Destination node D receives two RREQs from node 6 and node 4 respectively. D buffers both the paths for the route reply. Figure 3 shows another example of node disjoint path selection. Destination node D receives the two RREQ from node 4 and node 6 respectively. Since both the routes shares a common node i.e. node 1, destination node D chooses route based on greater value of Thresh_Delay in RREQ.

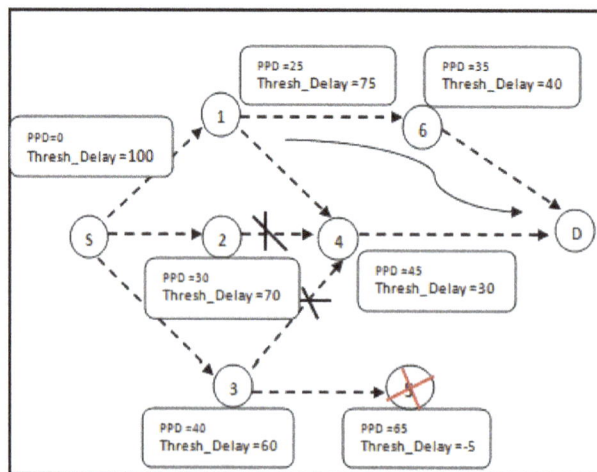

Fig. 3: Route acceptance by Destination

2.3. ROUTE REPLY

In proposed CLDM-AODV destination node D can collect up to RREQcounter times RREQ packets within time duration RREQ_WAIT_TIME, which is assumed to 20 ms. Node D generates a route reply RREP packets in response to every RREQ copy that arrives from the source S via loop-free and node disjoint paths to the destination. RREP packet is an extension of AODV RREP packet with additional field Max_PPD, which will hold the maximum packet processing time at intermediate nodes along the reverse path. Before destination node forwards the RREP, it computes the PPD and updates it in the Max_PPD field as shown in Figure 4. On reaching the next node, the intermediate node computes its PPD and compares it with the value in the Max_PPD field of RREP packet if current PPD computed is more than value in the Max_PPD. On receiving the RREP from all the disjoint routes, the source selects the primary route with minimum Max_PPD value. This signifies that the packet travelled through the less congested network, and possibility of packet incurring extra delay or getting dropped on the path is very low. Figure 5 shows an example of node disjoint route reply procedure. Destination node D calculates its PPD which is 25 ms and initializes Max_PPD with that PPD. Node D then sends RREP packets to all QoS qualified RREQ routes. Intermediate nodes 4 and 6, on receiving the RREP compute their own PPD i.e. 45 ms and 15 ms respectively. This value is compared with Max_PPD field of RREP packet. If PPD value is less or equal to Max_PPD, it ignores else it replaces the Max_PPD value in RREP packet. Node 6 does not modify Max_PPD as its computed PPD value is less than Max_PPD whereas node 4 replaces Max_PPD with 45 ms as its computed PPD value is greater than Max_PPD. Source node S on receiving the multiple RREP, it buffers them in the route table. Source S chooses the path with minimum value of Max_PPD as primary path i.e. path which source receives from node 1 as its Max_PPD value is 25 ms. If source does not receive RREP in RREP WAIT_TIME from destination, then it restart route discovery with new session Id.

2.4. ROUTE MAINTENANCE.

Route maintenance is very essential as there are high chances of route failure and QoS constraint violation due to mobility. Route failure due to link breakage is handled by the method using periodic *Hello* packets [15]. Any node which detects either a QoS violation or a link failure, informs the source by sending a route error packet (RERR). If a source node itself moves, restart the route discovery procedure to find a new route to the destination. If a node along the route moves so that it is no longer reachable, its upstream neighbor sends a link failure notification message to each of its active upstream neighbors through RERR until reaches the source node. QoS violation due to end-to-end delay constraint is detected by the intermediate nodes by computing one way delay experienced by the data packets from the sender's timestamp on the received data packets. During data transmission, source node appends the Thresh_Delay information to the data packets. Intermediate nodes on receiving the data packets, finds the difference between current time and time stamp of data packet. If value is less than Thresh_Delay, it generates the RERR packet to the source, or else forwards the packet to the next hop in the route table.

In our proposed CLDM-AODV, we introduce a method to validate other alternate node disjoint paths already discovered. At regular interval of time, Life Line Packets (LLP) is forwarded through alternate paths which contain Thresh_Delay. Intermediate nodes on receiving LLP, verifies the eligibility of packet forwarding by computing difference between current time and time stamp of data packet. If it is less than Thresh_Delay, it generates the RERR packet to the source indicating that the path is no longer QoS compliance link and corresponding path entry is deleted from route table. Destination node replies to these LLP by the same procedure as followed during RREP packets. On receiving the fresh route quality, source updates the primary path with highest quality.

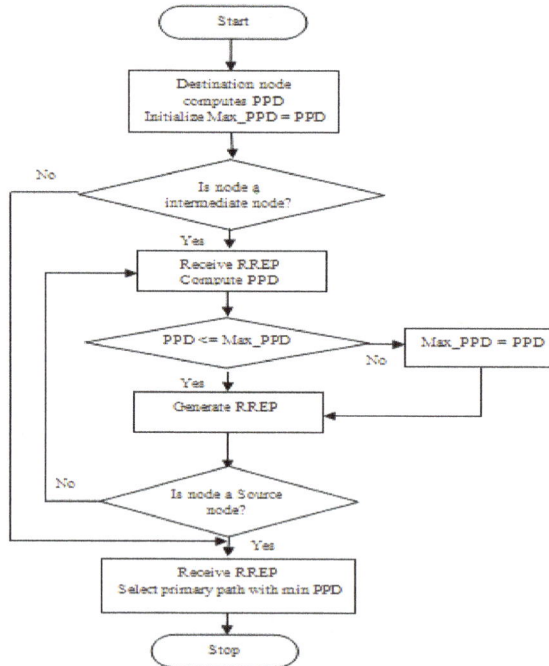

Fig. 4: RREP Flowchart of proposed CLDM-AODV

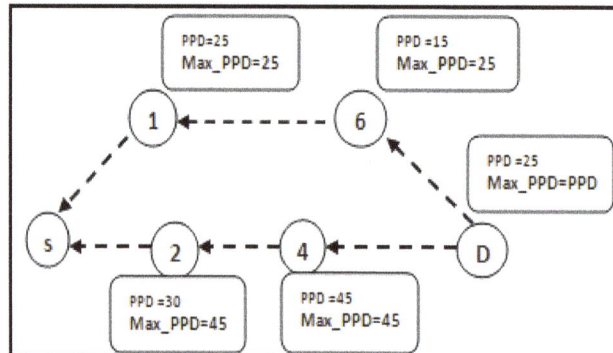

Fig. 5: Route Reply of proposed CLDM-AODV

3. SIMULATION EXPERIMENTS

3.1 Simulation Environment

The performance of the proposed CLDM-AODV protocol is evaluated and compared with AODV and NDMR. Simulations are conducted on the Network Simulator (ns-2) with network comprising of 50 wireless ad hoc nodes moving over an area of 1500m x 300m for 900s of simulated time. Physical layer is a bi-directional link and channel transmission rate is 2Mbps. At MAC layer, the DCF of IEEE 802.11 standard for wireless LANs is assumed. RTS and CTS packets are exchanged before the transmission of data packets. The channel propagation model we used two-ray ground reflection model. Constant Bit Rate (CBR) traffic is used. A 512-byte data packet with 2 packets/second sending rate is assumed for all the experiments. Inter packet time is assumed to be 35 ms. Radio transmission range of each node is set to 250m. The initial placement of nodes is random and random waypoint mobility model [24] is used to simulate node movements. Simulation is run for seed value of 1 to 9.

The simulation parameters are shown in table 1.

Table 1

Parameters	Value
NS version	Ns –allinone-2.35
Number of nodes	50
Simulation Time	900 sec
Radio transmission range	250m
Traffic	CBR(Constant Bit Rate)
CBR Packet size	512 bytes
Simulation Area size	1500m * 300 m
Node Speed	4m/s to 20 m/s
Mobility model	Random WayPoint mobility

We compare the performance of AODV, NDMR and CLDM-AODV using the following three metrics:

1. *Control Overhead* is the ratio of the number of protocol control packets transmitted to the number of data packets received.

2. *Packet Delivery Fractions (PDF)* is the ratio of the data packets delivered to the destination to those generated by the CBR sources.

3. *Average end-to-end delay* is an average end to delay of all successfully transmitted data packets from source to destination.

3.2 Simulation Results

Example 1: In this example, we analyze the effect of speed on control overhead, PDF and average end to end delay for different number of source nodes in the network. In the simulation we assume the number of sources to be 30, 35 and 40 and mobility of nodes is 4 meters/sec to 20 meters/sec.

Figure 6(a)-(c) shows the plot of control overhead vs. speed. It is evident from the result that CLDM_AODV has minimum control overhead compared to AODV and NDMR. In Figure 7, average control overhead ratio for sources 30, 35 and 45 is plotted. It is easily inferred that CLDM_AODV has smaller overhead than AODV and NDMR in harsh operation environments. This improvement is mainly because multiple QOS compliance routes are discovered in single route discovery phase, which significantly reduces frequent route discovery on route failure.

Figure 8(a)-(c) shows the plot for End-to End delay vs. speed. It can be seen from the plot corresponding to AODV that there is an increase in delay which is due to high mobility of nodes which in turn results in increased probability of link failure that causes an increase in the number of routing rediscovery processes. This makes data packets to wait for more time in its queue until a new routing path is found. Average end-to-end delay in NDMR does not show much variation over varying speed and shows better results compared to AODV.

In Figure 9, average End-to End delay vs. speed for sources 30, 35 and 45 is plotted. In proposed CLDM-AODV protocol, delay curve remains consistently low compared to AODV and NDMR even though extra waiting time, RREQ_WAIT_TIME, is added in route discovery process. Addition of RREQ_WAIT_TIME has little effect on the overall performance since CLDM-AODV has multiple alternate node disjoint paths satisfying the delay constraint, leads to less route discoveries. Also source regularly uses the primary path with optimal quality.

A packet delivery ratio for AODV, NDMR and CLDM_AODV is as shown in Figure 10(a)-(c). In Figure 11, average Packet delivery ratio vs. speed for sources 30, 35 and 45 is plotted. Since CLDM_AODV attempts to use optimal QoS enabled node disjoint path among available multiple alternate paths for data delivery, the protocol is able to deliver more packets to the destination compared to AODV and NDMR.

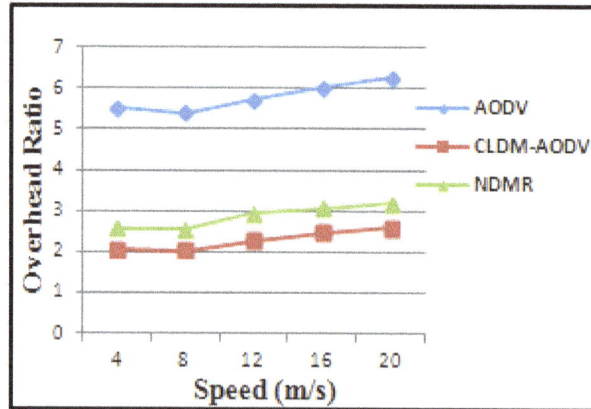

(a)   30 source nodes

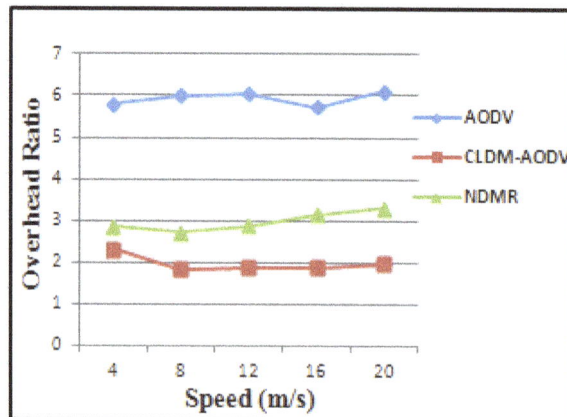

(b)   35 source nodes

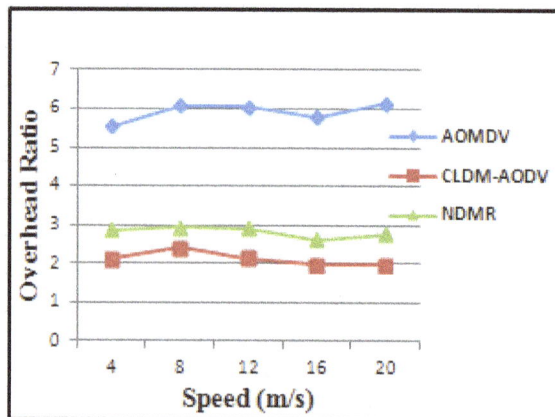

(c) 40 source nodes

Fig. 6(a)-(c): control packet overhead vs. speed (m/s).

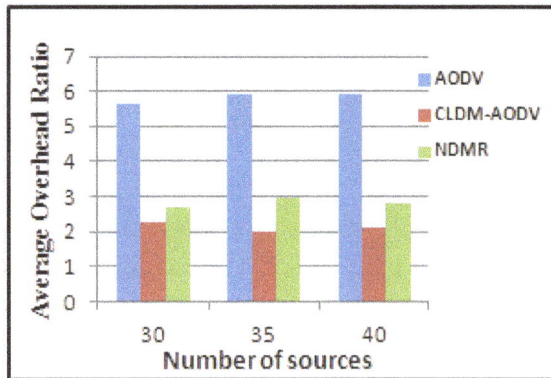

Fig. 7: Average Control overheads for varying number of sources for speed 4m/s to 20 m/s

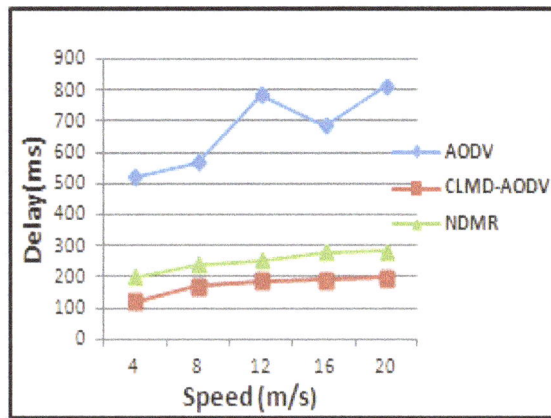

(a) 30 source nodes

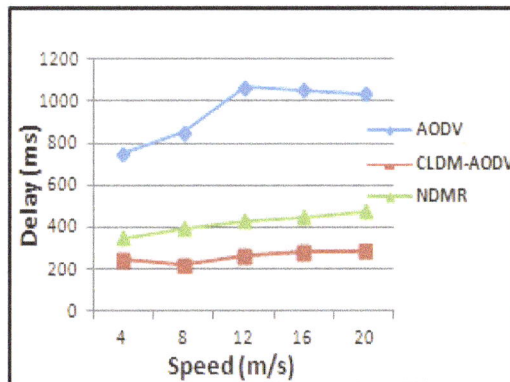

(b)   35 source nodes

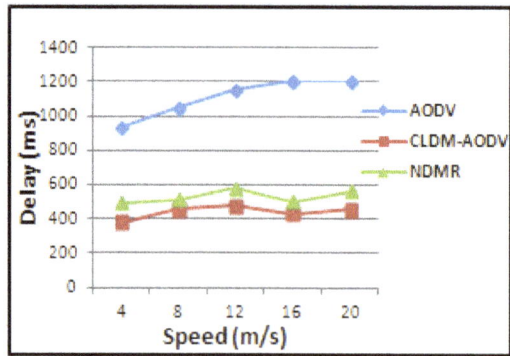

(c) 40 source nodes

Fig. 8(a)-(c): End-to-End delay (ms) vs. speed (m/s).

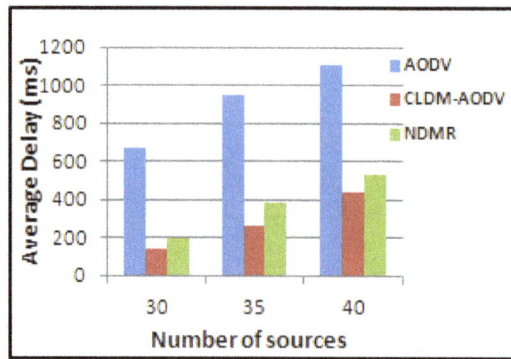

Fig. 9: Average end-to-end for varying number of sources from speed 4m/s to 20 m/s

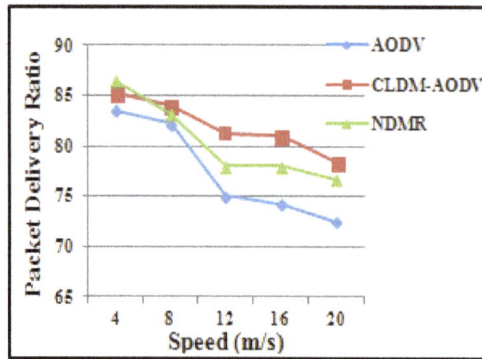

(a) 30 source nodes

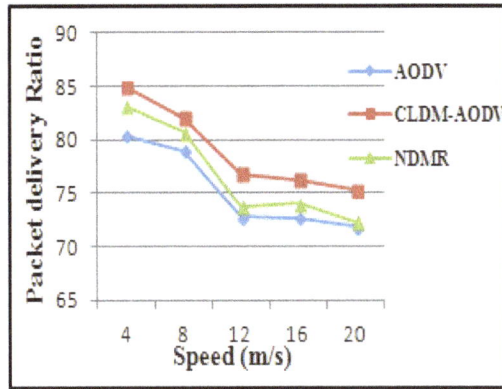

(b) 35 source nodes

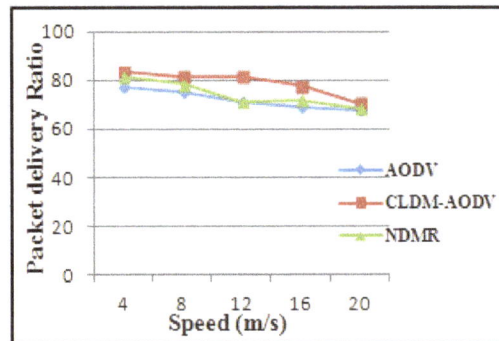

(c)40 source nodes

Fig. 10(a)-(c): Packet delivery ratio vs. speed (m/s).

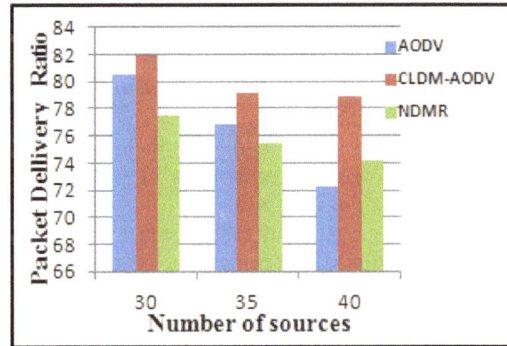

Fig. 11: Average Packet delivery ratio for varying number of sources from speed 4m/s to 20 m/s

AODV simply drops data packets when routes are disconnected, as it has to resort to a new discovery when the only path fails. Proposed CLDM_AODV algorithm performs better as data packets travel through less congested and delay compliance.

IV. CONCLUSION

A new algorithm CLDM-AODV suitable for delay sensitive application is presented. Proposed CLDM-AODV algorithm with multipath capability effectively deals with high mobility traffic route failures in MANET. Proposed algorithm ensures that the multiple paths are loop-free and is node disjoint. Comparative study of CLDM-AODV, classical AODV and NDMR is performed using ns-2 simulations under varying mobility and traffic scenarios. The results indicate that CLDM-AODV has lower average end-to-end delay even by including extra fields to RREQ and RREP packets to provide QOS support. The routing overhead is low compared to its counter parts as route discovery process is minimized by providing QOS compliance alternate routes. The added advantage of the proposed algorithm is, it periodically checks the paths obtained during route discovery process and uses optimal link for data communication.

REFERENCES

[1]   R. Jurdak, Wireless Ad Hoc and Sensor Networks 9th ed. Springer Series, United States of America.2006.

[2]   C.K.Toh, Ad Hoc Mobile Wireless Networks: Protocols and Systems. 2nd ed. Printice Hall, China, 2001.

[3]   C. E. Perkins, E. M. Belding-Royer, "Quality of Service for Ad hoc On- Demand Distance Vector Routing, Internet Draft," October, 2003.

[4]   X. Li and L Cuthbert, "Multipath QoS routing of supporting Diffserv in Mobile Ad hoc Networks," Proceedings of SNPD/SAWN'05, 2005.

[5]   Z. Ye, S. V. Krishnamurthy and S. K. Tripathi, "Framework for Reliable Routing in Mobile Ad Hoc Networks," IEEE INFOCOM 2003.

[6]   P. Sambasivam, A. Murthy, and E. M. Belding-Royer, "Dynamically Adaptive Multipath Routing based on AODV," proc of the 3rd Annual Mediterranean Ad Hoc Networking Workshop (MedHocNet), Bodrum, Turkey, June 2004.

[7]   P. Macharla, R. Kumar, A. Kumar Sarje, "A QoS routing protocol for delay-sensitive applications in mobile ad hoc networks," COMSWARE. 2008, pp. 720-727.

[8]   S.Chakrabarti, and A. Mishra, "QoS issues in ad-hoc wireless networks," IEEE Communication Magazine, Feb 2001,Vol. 39, No. 2, pp.142 148.

[9]   M.K. Gulati and , K. Kumar, "A review of QoS routing protocols in MANETs", 2013 International Conference on Computer Communication and Informatics (ICCCI -2013), Jan. 4 - 6, 2013.

[10] C. Chen, W. Wu Z. Li, "Multipath Routing Modeling in Ad Hoc Networks," Proc. of IEEE ICC, May 2005, pp.2974-2978.

[11] P. Wannawilai, C. Sathitwiriyawong, "AOMDV with Sufficient Bandwidth Aware," Computer and Information Technology (CIT), 2010 IEEE 10th International Conference on Computer and Information Technology, June 2010, pp.305-312.

[12] C. Ahn, S. Chung, T. Kim, and S. Kang, "A node-disjoint multipath routing protocol based on aodv in mobile ad hoc networks," In Information Technology: New Generations (ITNG), 2010 Seventh International Conference, April 2010, pp. 828 –833.

[13] M.K.Marina , and S.R.Das, "On-Demand multipath distance vector routing in ad hoc networks," Proceedings of the 9th IEEE International Conference on Network Protocols (ICNP), 2001.

[14] C. E. Perkins, E. M. Royer, and S. R. Das, Ad hoc on-demand distance vector routing, Internet Draft, 2002.

[15] C.E. Perkins, and E.M. Belding-Royer, "Quality of Service for Ad Hoc on Demand Distance Vector Routing," draft-perkins-manet-aodvqos-02.txt, Mobile Ad Hoc Networking Working Group Internet Draft, 14 October 2003.

[16] S. Lohier, S. Senouci, Y. M. Ghamri Doudane and G. Pujolle, "A reactive QoS Routing Protocol for Ad Hoc Networks," European Symposium on Ambient Intelligence (EUSAI'2003), Eindhoven, Netherlands, November 2003.

[17] J. Park, S. Moh†, and I. Chung, "A Multipath AODV Routing Protocol in Mobile Ad Hoc Networks with SINR-Based Route Selection," ISWCS '08. Wireless Communication Systems, October 2008, pp. 682–686.

[18] E. Mehdi, E.MohammadReza, D.Amir, Z.Mehdi, and Y. Nasser, "Load Balancing and Route Stability in Mobile Ad Hoc Networks base on AODV Protocol," Proceeding of International Conference on Electronic Devices, Systems and Applications(ICEDSA2010), 11-14 April 2010, pp. 258 – 263.

[19] X. Li and L. Cuthbert, "Multipath QoS Routing of supporting DiffServ in Mobile Ad hoc Networks," SNPD-SAWN '05 Proceedings of the Sixth International Conference on Software Engineering, IEEE Computer Society Washington, DC, USA 2005, pp. 308-313.

[20] Boshoff, J.N., Helberg, A.S.J. (2008), " Improving QoS for Real-time Multimedia Traffic in Ad-hoc Networks with Delay Aware Multi-path Routing ," IEEE, Wireless Telecommunication Symposium, WTS 2008, pp. 1-8.

[21] P. Sambasivam, A. Murthy, E. M. Belding-Royer, "Dynamically Adaptive Multipath Routing based on AODV," Med-Hoc- Net, (2004), pp. 16-28.

[22] M. Obaidat, "A Novel Multipath Routing Protocol for Manets," Wireless Communications, Networking and Mobile Computing (WiCOM), 2011, pp.1-6.

10

PROBABILISTIC ROUTING USING QUEUING THEORY FOR MANETs

Gaurav Khandelwal, Giridhar Prasanna, Chittaranjan Hota

Birla Institute of Technology & Science-Pilani
Hyderabad Campus, Hyderabad, INDIA
{gaurav, prasanna, hota}@bits-hyderabad.ac.in

Abstract

Mobile ad-hoc networks pose real difficulty in finding the multihop shortest paths because of continuous changing positions of the nodes. Traditional ad-hoc routing protocols are proposed to find multi-hop routes based on shortest path routing algorithms, which cannot effectively adapt to time-varying radio links and network topologies of Ad-hoc networks. In this paper we proposed an enhanced routing algorithm, which uses probabilistic approach for the stability of the neighboring nodes in finding and maintaining the routing paths in Ad-hoc networks. The probability of a node being stable in the path is modeled by queuing theory, where the stability of a node is measured by number of packets arrived at a node and the number of packets being serviced by the node per unit time. Proposed approach shows significant improvement over the traditional Ad-hoc on-demand distance vector routing protocol as analyzed in the result analysis section.

Keywords

AODV, MANET, Probabilistic routing, Stability, Queuing theory.

1. INTRODUCTION

An Ad-hoc wireless network consists of a group of mobile nodes which communicate with each other through the wireless links in a distributed fashion without a centralized controller. Mobile Ad-hoc Networks (MANETs) are autonomous collection of mobile nodes. MANET usually lacks any type of fixed infrastructure. As nodes in MANET are mobile, so the link failure and re-establishment of routes takes place frequently. It is difficult to keep track of all these nodes and their locality centrally. So it is essential to find out the geographic location of nodes only when communication channel has to be established, in this wireless environment wherein bandwidth available for the communication is constrained. Moreover, these nodes keep moving with different speed and power is an important constraint for them all the times. A topology of 18 nodes Ad-hoc network is shown in figure 1. Because the nodes are mobile, the topology keeps changing which makes the transfer of packets complicated.

To discover and maintain the routes in ad-hoc networks requires more control traffic which makes the task of ad-hoc routing more complex and less efficient. Several researches on developing ad-hoc routing protocols for MANETs have been proposed [1, 5]. These routing protocols can be classified as proactive protocols [1, 2], reactive protocols [4], or hybrid of the both [5].Pro-active (table-driven) routing protocols maintain fresh lists of destinations and their routes by periodically distributing routing tables throughout the network. The main disadvantages of such algorithms are the respective amount of data for maintenance and slow reaction on restructuring and failures. Example of pro-active algorithm is Wireless Routing Protocol (WRP) [7] which uses an enhanced version of the distance-vector routing protocol that uses the classical Bellman-Ford algorithm [9] to calculate paths. Reactive (on-demand) routing protocols finds a route on demand by flooding the network with route request packets. The main disadvantages of such algorithms are high latency time in route finding and excessive flooding can lead to network clogging. Examples of reactive algorithms are Dynamic Source Routing (DSR)[7] and Ad-hoc On-demand Distance Vector (AODV)[6]. AODV is designed for networks with tens to thousands of mobile nodes. One feature of AODV is the use of a destination sequence number for each routing table entry. The sequence number is created by the destination node. The sequence number included in a route request or route reply is sent to requesting nodes. Sequence numbers are very important because they ensure freedom from loops and are simple to program. Sequence numbers are used by other nodes to determine the freshness of routing information. If a node has the choice between two routes to a destination, a node is required to select the one with the greatest sequence number. In AODV, every node has a routing table. In AODV, when a source node needs a connection to a destination, it broadcasts a route request which is forwarded by all other nodes in the path. Intermediate nodes record the node from which they have received this request which creates an explosion of temporary routes back to the source node. When an intermediate node receives such a message and already has a route to the desired node, it sends a message backwards through a temporary route to the requesting node. The source node then begins using the route that has the least number of hops through other nodes. It's entries are destination IP address, prefix size, destination sequence number, next hop IP address, lifetime (expiration or deletion time of the route), hop count (number of hops to reach the destination), network interface, other state and routing flags (e.g. valid, invalid). Route Requests (RREQs), Route Replies (RREPs) and Route Errors (RERRs) are message types defined in AODV. When a link fails in AODV, a routing error is passed back to a transmitting node, and the process repeats.

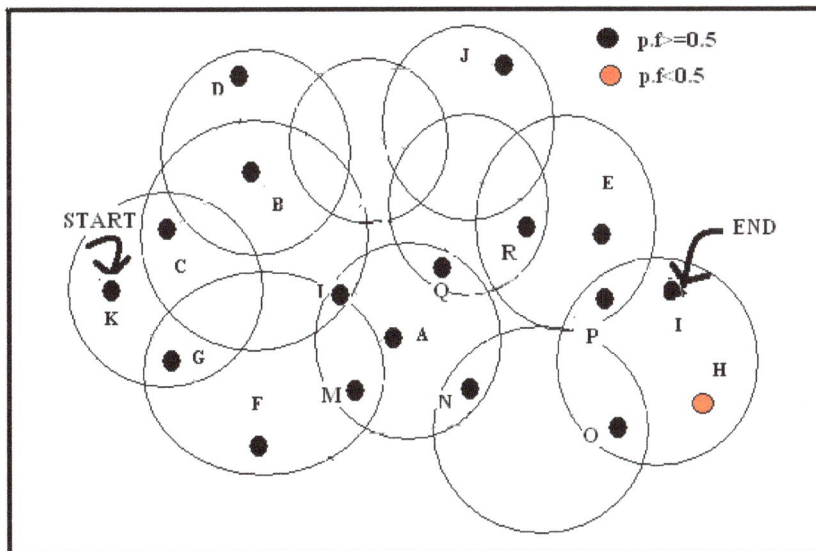

Figure 1. Topology of nodes in an Ad-hoc network

Due to the dynamic changes of network topology in MANETs, however, such a protocol must constantly acquire new information to exactly reflect the network topology up-to-date, which will trigger a high overhead, especially in the presence of high mobility and large number of nodes. Also, it significantly limits its practicability [6, 9]. Therefore, the available deterministic protocols can carry out the routing process only under a group of restricted constraints on the network coverage area size, node density or mobility. Probabilistic protocols, on the other hand, approximately predict the network topology using only probabilistic and traceable information [11, 16]. Therefore, they can significantly reduce the amount of overhead in comparison with the deterministic protocols and thus are more scalable and topology independent.

In this paper, we propose an enhanced probabilistic routing protocol (PAODV) by introducing a queuing model for infinite packet arrivals at a node, such that the new protocol can work efficiently under different network characteristics like node density, coverage area size, and node mobility. Although we choose PAODV routing protocol to guarantee node mobility, resiliency and low latency, the basic idea of this new method can be applied to other probabilistic gossip-based routing protocols. Mathematical and extensive simulation analyses have been performed to verify the efficiency of the enhanced routing scheme in coping with different network sizes and topologies as well as node mobility. Each experiment was inspected in terms of packet delivery probability, packet drop, and throughput which have been reported in later sections.

The remainder of the paper is organized as follows: Section 2 reviews the related research in MANET routing algorithms. In Section 3, we discuss assumptions and the proposed algorithms. In Section 4, we present the simulation run. Section 5 analyzes the results obtained from simulation runs. Finally, Section 6 concludes the paper and presents the future work.

2. RELATED WORK

Yassein et al. [29] proposed flooding which is static routing protocol which does not need any information about the network topology to deliver packets from the source to the destination. When a node wants to send a packet, it transmits this packet to all its neighbors. Then each node that receives this packet for the first time retransmits it to all neighbors except the neighbor from which it has received which is continued until the packet reaches all nodes in the network. Each packet has a unique identifier that consists of the source address, a special sequence number used to prevent sending duplicate packets from the same node, and the destination address. The main disadvantage of flooding is the consumption of the network resources because of the high traffic load it generates. On the other hand, it ensures that the packet reaches the desired destination and gives a high packet delivery ratio.

Reactive routing schemes were developed to reduce routing overhead in mobile ad-hoc networks. Today, reactive (or on demand) routing protocols have become synonymous with the flooding of route requests (RREQs) when a path needs to be established. While this approach may be the fastest solution in a network that is not bandwidth-limited, it leads to the broadcast storm problem as identified by Ni et al. [34], especially in volatile routing environments. This inefficiency has been identified by many researchers in the past, and several optimizations over this blind flooding have been proposed. These approaches include the use of an expanding ring search, the use of heuristics based on connected dominating sets to reduce the number of nodes retransmitting the packets [35], the use of geographical information to direct the flooding [36] and probabilistically reducing the number of retransmissions [34].

There are many protocols that implement solutions to the flooding problem in on-demand routing protocols by enhancing route recovery mechanisms [30][31]. In general, routing protocols in MANETs can be classified into deterministic and probabilistic protocols. Some deterministic approaches try to enhance the existing deterministic protocols to get more scalable algorithms. SMORT [17] is a scalable deterministic routing algorithm that exploits secondary paths to recover from broken paths, and thus reduces the overhead produced in route recovery procedure of AODV protocol. In presence of only a few sessions, this protocol provides good scalability for different sizes of the network by adopting fail-safe multiple paths. Ad-hoc On-demand Stability Vector (AOSV) routing protocol [20], is proposed to properly and effectively discover stable routes with high data throughput and long lifetime by considering the radio propagation effect on signal strength. Some also use Global Positioning System (GPS) [21] to improve the AODV by calculating reliable distance. In [22], authors have proposed a stable, weight-based, on-demand routing protocol.While the proposed scheme may combat against link breaks due to mobility, link breaks due to the draining node energy is a factor that also must be accounted for when computing weights for stable routing.

In AODV with Backup Routing (AODV-BR) [31], nodes overhear route reply messages of their neighbors to create their own alternate routes to destination. When a node detects a broken route, it broadcasts the packet to its neighbors hopefully that one of them has a valid route to the destination and at the same time sends a RERR message to the source to initiate a route rediscovery. The reason for reconstructing a new route instead of continuously using the alternate path is to build a fresh and optimal route that reflects the current status of the network. AODV-BR concentrates on increasing route reliability by decreasing packet drop rates but it suffers from two main problems: stale routes and duplicate packet transmission. An improvement of AODV protocol based on backup route (AODV-BR) in mobile ad hoc networks was proposed by Lee et al. [25]. AODV-BR establishes the mesh and multi-paths to the destination. In AODV-BR, the primary route and alternate routes together establish a mesh structure that look similar to a fish bone. AODV-BR increased PDR but, has longer end-to-end delay.

In Neighborhood-aware Source Routing (NSR) protocol [32], each node has a partial topology that covers in addition to the 2-hop neighborhood, the links in requested paths to destinations. Link state information is maintained by broadcasting periodic HELLO messages. In case of route failure, an intermediate node tries to repair the route if either the link to the next hop has failed or the link headed by the next hop on the path to be traversed has failed. RERR message is propagated to the source node if an intermediate node uses a completely new route to destination or it has no alternate route to destination. HELLO messages in NSR incur excessive overhead to maintain the partial topology of the network. Additionally, stale route problem may affect the performance of NSR.

In AODV with Backup Routing (AODV-BR) [31], nodes overhear route reply messages of their neighbors to create their own alternate routes to the destination. When a node detects a broken route, it broadcasts the packet to its neighbors hoping that one of them has a valid route to the destination and at the same time sends a RERR message to the source to initiate a route rediscovery. The reason for reconstructing a new route instead of continuously using the alternate path is to build a fresh and optimal route that reflects the current status of the network. AODV-BR concentrates on increasing route reliability by decreasing packet drop rates but it suffers from two main problems: stale routes and duplicate packet transmission.

Jian et al. [26] proposed an improvement on AODV based on reliable delivery (AODV-RD) in mobile Ad-hoc networks. In AODV-RD, a link failure fore-warning mechanism, metric of alternate node in order to better select, and also repairing action after primary route breaks is performed. AODV-RD is an improvement in AODV-BR. In this work, link failure prediction mechanism is used for checking the strength of the packet signal [27]. If the strength is in warning state then an alternate route is discovered. If more than one alternate paths are present with same hop count then the route is selected whose communicating power is stronger. AODV-RD significantly increased packet delivery ratio (PDR).

Sethi et al. [28] proposed an Optimized Reliable Ad-hoc On-Demand Distance Vector (ORAODV) protocol that offers quick adoption to dynamic link conditions, low processing and low network utilization in Ad-hoc networks. They used a mechanism of retransmission of undelivered data packets with blocking technique to enable optimal path routing and fast route delivery with an improvement of PDR.

Yassein et al. [33] proposed a Smart Probabilistic Broadcasting (SPB) scheme as a new probabilistic method to improve the performance of existing on-demand routing protocols by reducing the RREQ overhead during the rout discovery operation. The simulation results showed that the combination of AODV and a suitable probabilistic route discovery can reduce the average end-to-end delay as well as overhead, while achieving low normalized routing load, compared to AODV that uses fixed probability and blind flooding.

In [23], authors presented efficient broadcasting schemes that combine the advantages of pure probabilistic and counter-based schemes. The re-broadcast decision depends on both fixed counter threshold and forwarding probability values. The value of probability is set according to packet counter that is not equal to the number of node neighbors. In [24], the authors introduced a technique to reduce the RREQ overhead during route discovery operation, using the previous path. They argue that when the path between source and destination is changed, the new path between them will not be extremely different from the previous one. Beraldi et al. [10] presented a preliminary idea of hint-based routing for Ad-Hoc networks exploiting the duration of time passed since the last time nodes encountered with the destination, namely the encounter age. Zone Routing Protocol (ZRP) [5] is another technique that was proposed to reduce RREQ control packets, which uses a combination of two protocols, namely proactive and reactive; it takes the advantages of both protocols in order to solve the flooding of RREQ control packets.

3. PROPOSED ALGORITHM

From the findings of the birth and death process in queuing model [8], we can relate our proposed algorithm to the First Come First Serve (FCFS) model. The arriving packets can be regarded as the arriving customers in an unbounded queue and the route finding algorithm as the service provided to these customers (packets). According to the FCFS model, the traffic arrives randomly, which can be treated as a Poisson distribution, and are served in FCFS basis, which is an exponential distribution. Thus the probability that there will exist 'n' or greater than 'n' number of packets at a given time in a system is given by:

$$P(n) = (1-\mu/\beta)(\mu/\beta)^n \dots\dots\dots\dots\dots\dots\dots\dots\text{Eq. (1)}$$

Where, 'μ' is the arrival rate of the packets at a node, 'β' is the service rate of a node, and 'n' is the total number of nodes the packet has travelled, before the next node.

Thus in order to decide whether a packet at a node should be forwarded to another node on a particular route is decided by considering the strength of that route, which in turn in decided by the probability function described above in equation 1. Greater the strength of a particular route less will be the chances of packets being lost or dropped as that route is being continuously used to serve the packets for some period of time without any jitters.

The working of the proposed protocol is explained through the following example. Consider the topology of the MANET as shown in figure 1. Let the source and the destination nodes be K and I respectively. Let node H moves out from its current position, and the probability factor (p.f) of node H be less than 0.5. Probability of each node is computed using equation 1. In the route discovery phase, node K broadcasts RREQ messages to node C and node G. Now, node C and node G will broadcast RREQ messages to node B and node F. To the HELLO messages from node B and F, only nodes M, L and D. But the node D has a broken link, so it won't broadcast further. In the similar manner, the RREQ from node M and node L will reach node A. Node A will further broadcast to Q and N which further broadcast to node R and O respectively.Node O broadcast to node H and node R will broadcast to E.Node H would not broadcast as it's probability is less than p.f. . Node E will broadcast to P which will broadcast to I,which is our final destination. RREPs(route reply) will go through the shortest path from node I to node P to node E and than finding the shortest path from node E to node K through node Q,node A,node L,node B,node C. Without the proposed approach, AODV would have selected the path as K->G->F->M->A->N->O->H->I (as it is the shortest path). As shown in figure 2,node H moves out from the network topology, so there is no path from H to I and hence route discovery step is again carried out which is not the case in the proposed algorithm. Proposed approach takes care of the probability of node H moving out of the range of node F which in turn would have resulted in packet drop and finding a new path again. This in turn would have increased the network traffic in AODV which is avoided in the proposed approach.

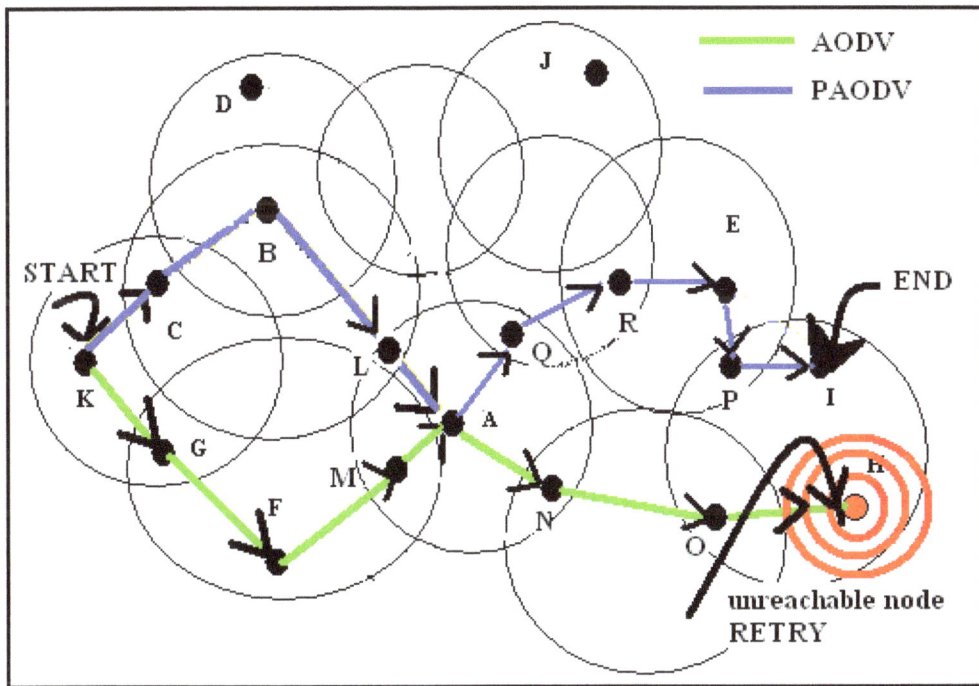

Figure 2. An example path using PAODV

In this algorithm, if the probability of a route is less than the threshold value p.f (which we have assumed to be 0.5 i.e. 50% stronger route), then another route for forwarding the path is determined. The arrival of packets is calculated by the difference between current time and time when the previous packet arrived. The service time is assumed to be a constant rate of 0.001 seconds per packet. Figure 3 shows the flow chart of the proposed algorithm. Comparison of performance evaluation of AODV [6] and the proposed algorithm PAODV is shown in Table 1 where 'n' is the number of nodes in the network and 'd' is the network diameter. Complexity of AODV is obtained from the work on routing protocols for ad-hoc networks [3].

Table 1. PERFORMANCE EVALUATION

PERFORMANCE PARAMETERS	AODV	PAODV
Time Complexity (initialization)	$O(2d)$	$O(2d)$
Time Complexity (post failure)	$O(2d)$	$\leq O(2d)$
Communication Complexity (initialization)	$O(2n)$	$O(2n)$
Communication Complexity (post failure)	$O(2n)$	$\leq O(2n)$

Algorithm 1: Finding different paths based on probability	**Algorithm 2:** Neighbor availability check

function PAODV_rreq
    nowtime = GetCurrentTime();
    pkt = queue.get();
    while(pkt)
        stamp = nowtime;
        ratio = arrival/β;
        factor = ratio^n;
        probability = (1-ratio)*factor
        if((rreq_lifetime <= nowtime)
OR(prob<p.f))
            new_route = PAODV_rreq();
        else
            dequeue(queued_pkt);
            Pkt++;
        n++;
        time_curr = GetCurrentTime();
        arrival= time_curr – stamp;
        stamp = time_curr;
    end while
end
	function PAODV_checkneighborlist
	    nowtime = GetCurrentTime();
	    neighbor = nei.getlist();
	    while(neighbor)
	        stamp = nowtime;
	        ratio = arrival/β;
	        factor = ratio^n;
	        probability = (1-ratio)*factor
	        if((neighbor_time < nowtime)**OR**(prob<p.f))
	            unreach_list = new list();
	            Rt_entry = routetable_get(neighbor_addr);
	            if(Rt_entry)
	                unreach_list = dequeue(neighbour);
	                Rt_entry ++;
	        neighbor ++;
	    time_curr = GetCurrentTime();
	    arrival = time_curr – stamp;
	    stamp = time_curr;
	    n++;
	    end while
	end

The notations (variables) used in Algorithms 1 and 2 are described as below:
'n' is the number of nodes visited, 'arrival' is the arrival rate of packets, 'ratio' is equal to (arrival rate /β), and 'β' is the service rate which is equal to 0.001.

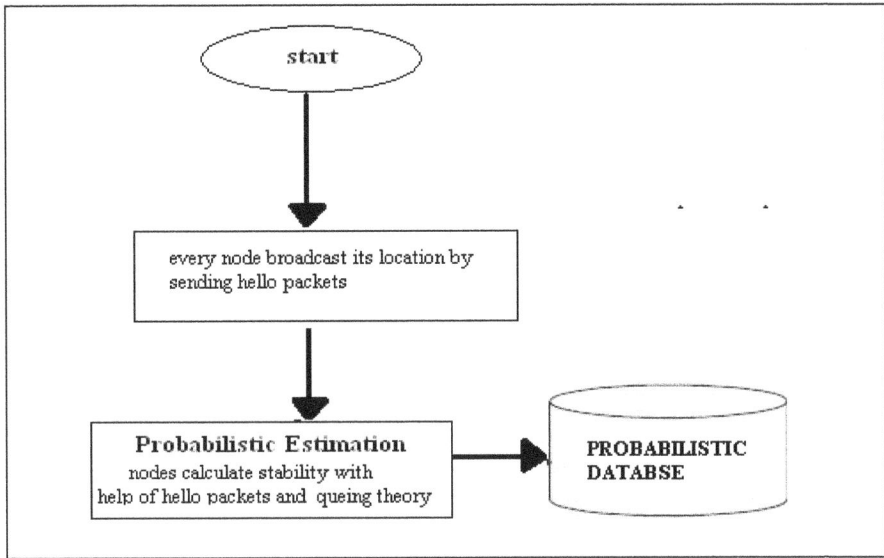

Figure 3a. Flowchart for computing probability of node stability

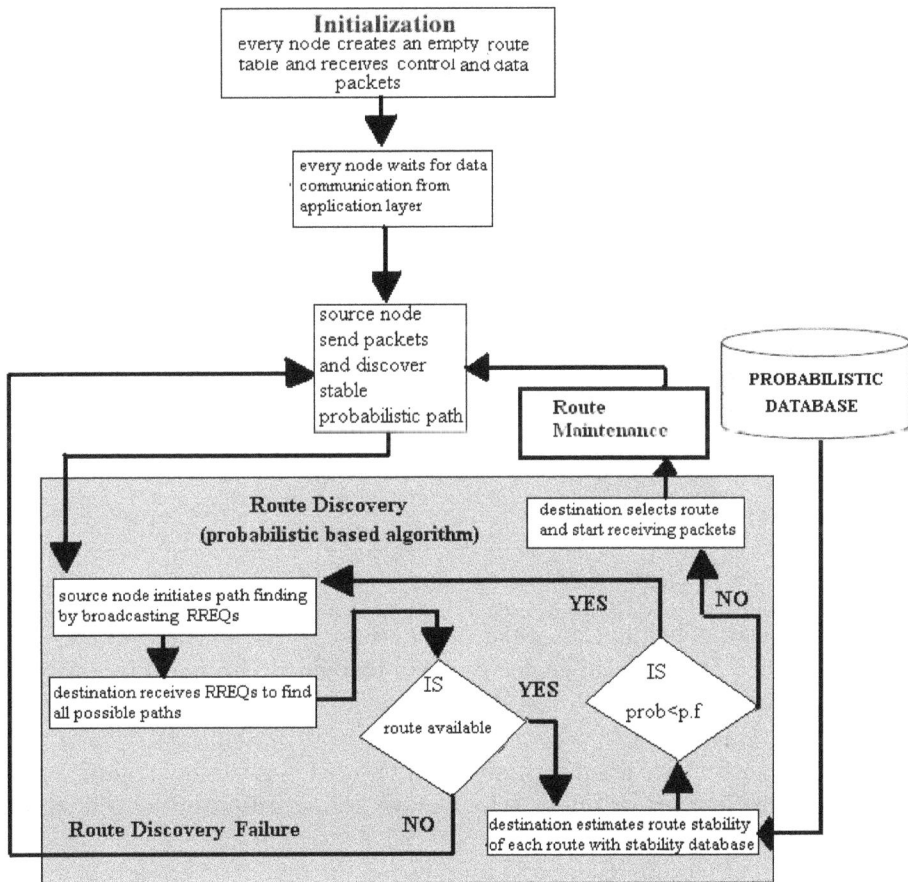

Figure 3b. Flowchart of proposed PAODV algorithm

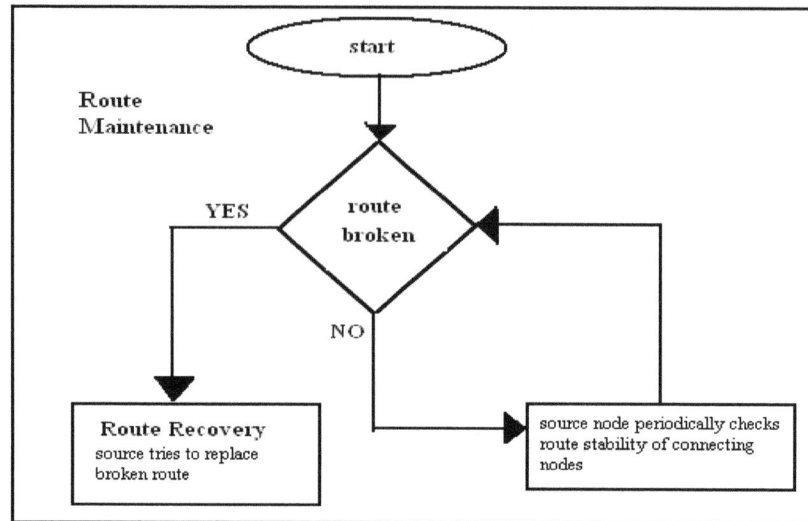

Figure 3c. Route maintenance in PAODV algorithm

Flowchart of the proposed PAODV is shown in figure 3a-3c. Figure 3a shows the flowchart of probabilistic database in which probabilistic estimation is made, which is used to compute the stability of the route as shown in figure 3b. Figure 3b shows route discovery phase of the proposed algorithm. In route discovery phase, route for packet transfer is computed using probabilistic database. Figure 3c shows route maintenance phase of the proposed PAODV. In this phase, it is checked whether the link is stable or broken, at regular interval of time.

4. SIMULATION

The simulation of the proposed approach was carried out in machines running Fedora 12 with Nctuns-6.0 as the network simulator [18, 19]. The simulation parameters are summarized in Table 2 as below:

Table 2. SIMULATION PARAMETERS

Simulation time	80 units
Phy-model	IEEE 802.11(b)
nodes	25,100,150
Node movement	Randomly generated
speed	10 m/s

NCTUns [18, 19] uses a distributed architecture to support remote and concurrent simulations. It uses open-system architecture to enable protocol modules to be added to the simulator. This task can be easily done in just a few mouse clicks via its GUI operating environment. The fully integrated GUI environment enables a user to edit a network topology, configure the protocol modules of a network node, set the parameter values of a protocol module, specify mobile nodes moving paths, plot performance curves, playback animations of logged packet transfers, etc. From a network topology perspective, the GUI program can generate a simulation job description file suite. Since the GUI program uses the Internet TCP/IP sockets to communicate with other components, it can submit a job to a remote simulation machine for execution. When the simulation is finished, the simulation results and generated log files are transferred back to the GUI. The user then either examines logged data, plots performance curves, or plays back packet transfer animations. We ran our simulations of Algorithm 1 and 2 using the parameters defined in Table 2, and the results were plotted. The results were plotted against varying number of network nodes. The traversal path of each node was generated randomly by using generate random path feature of the simulator. The simulations were carried out for 80 steps of the simulation time. To get realistic results, both the AODV [6] and the proposed PAODV were simulated with number of nodes varying from 25 to 100 to 150 (with a multiple of 5), having a total of 6 simulations. The packet size was fixed for all the simulations. Initially, the nodes were placed at random positions and then moved according to the randomly generated traversal path by the simulator. The moving speed of each node was kept constant throughout the simulation which was 10 m/s. This was done to ensure fairness across different simulations. By increasing the number of nodes in the topology and changing the probability factor of random nodes, better results were observed which are plotted in figure 4 to figure 7.

5. RESULT ANALYSIS

A. Packet drop as a Metric

Figure 4 shows the result of the packet drop when the simulation is done with the original AODV and with the proposed algorithm PAODV. The area under the curve gives the total amount of packet drop. The simulation is carried out for 80 seconds and for 18 nodes as shown in figure 1.When the area is calculated (area is calculated by trapezoidal rule) this simulation gives an improvement of about 18.6% i.e. there is 18.6% less packet drop of the PAODV algorithm as compared to the original AODV.

Figure 5 shows the simulation of 150 nodes for 80 seconds. With this simulation there is an improvement of nearly 26.8% i.e. packet drop is 26.8 % less in the proposed algorithm. As the number of nodes increase in the MANET, the improvement is also observed in this metric. Also, as the time of simulation increases, the improvement in the PAODV's performance is observed.

The reason for improvement can be explained easily because in the proposed algorithm, if the probability is less than p.f, then it will find a new path in that phase only rather than finding a new path again from starting as in AODV[6], which is described in flow chart of proposed algorithm shown in figure 3a-3c.

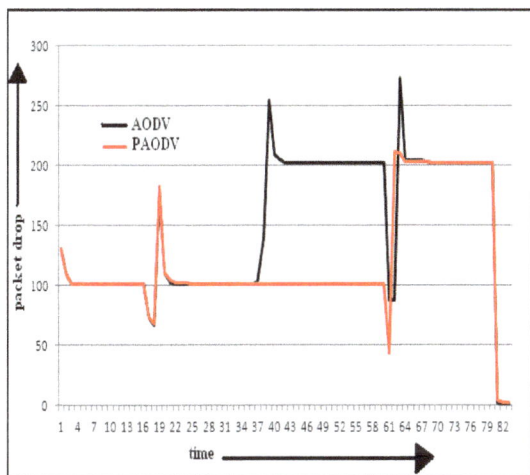

Figure 4. Time vs. number of packet drops

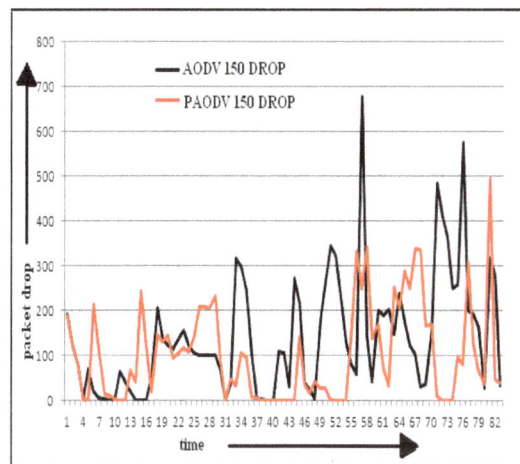

Figure 5. Time vs. number of packet drops

As the number of nodes increase, randomness increases and the problem moves towards more realistic state. There will be more alternate paths, and hence proposed algorithm chooses best possible path based on the probability and hence performance improves drastically as shown in figure 4 and figure 5.

B. Data throughput as a Metric

Figure 6 compares the result of the net-throughput of the original AODV with the proposed PAODV.As shown in figure 6; area under the graph of PAODV is more as compare with the original AODV. There is an improvement of nearly 3% for 25 nodes and nearly 22% for 150 nodes network.

C. Probability factor as a Metric

The probability factor (p.f) also changes the result improvement. A simulation of 100 nodes with proposed algorithm is carried out with p.f set to 0.8 and 0.5. Figure 7 shows the packet drop with p.f equal to 0.8 and 0.5.

The result shows that there is an improvement of 8.3% i.e. there is 8.3% less packet drop for p.f = 0.5 with respect to 0.8 p.f. The p.f = 0.5 is the optimal value and can be explained as follows:

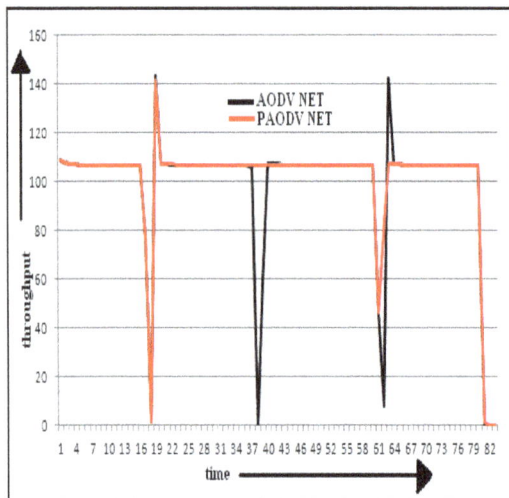

Figure 6.Time vs. throughput   Figure 7. Time vs. no. of packets drop

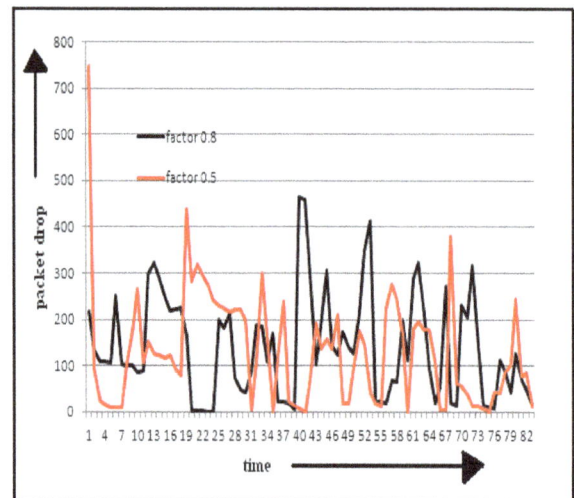

Because the value (0.5) is in between 0 and 1 and the improvement in our proposed algorithm is because it prevents the route propagation time when the node won't be available after route discovery. But this is the probabilistic approach and it might be possible that the node will be available for propagation which in turn wastes the time of route discovery again. The p.f when set to 0.5 keeps track of this and hence produces optimum result in general. Also, as the topology increases, the improvement is further observed.

6. CONCLUSION AND FUTURE WORK

In this paper we described a novel approach of using a probabilistic function to evaluate the packet drop in ad-hoc network. Our simulation results indicate that the packet drop reduces by around 27% for 150 node network and throughput increases by 21% for 150 node network. Therefore, we safely assume the proposed algorithm is a better approach in MANETs for ad-hoc on demand distance vector routing protocol(AODV) that are extensively used in the literature. It both optimizes the network performance and guarantees the communication quality. The proposed protocol provides a link failure fore-warning mechanism in the form of probability factor in order to better select an alternative route.

With simulation run of our algorithm, further properties of Ad-hoc networks may be measured. These include the in-broadcast, out-broadcast and comparisons of different nodes in the network rather than choosing some target node. As a continuation of this research in the future, we plan firstly to combine our algorithm with different routing protocols to observe, if a combined performance improvement is feasible. Secondly, we plan to study security threats along with probabilistic route finding, for secure transfer of data in MANETs. We also plan to improve upon end to end delay in data transfer.

REFERENCES

[1] C. Perkins and P. Bhagwat, "Highly dynamic destination-sequenced distance-vector routing (DSDV) for mobile computers", Proceedings of ACM SIGCOMM, pp. 234–244, (1994), London, UK.

[2] P. Jacquet, "Optimized link state routing protocol for ad hoc networks", Proceedings of IEEE International Multi Topic Conference, pp. 62– 68, (2001), Lahore.

[3] Elizabeth M. Royer, C-K Toh, "A Review of Current Routing Protocols for Ad-Hoc Mobile Wireless Networks", Proceedings of IEEE PersonalCommunications-99, pp 1-5(1999), Charleston, USA.

[4] D.B. Johnson, D.A. Maltz, and J. Broch, " DSR: The dynamic source routing protocol for multi-hop wireless ad hoc networks" , Ad Hoc Networking, ed. C.E. Perkins, Addison-Wesley, (2001).

[5] Z.J. Haas, M.R. Pearlman, and P. Samar, "The zone routing protocol (ZRP) for ad hoc networks", Internet draft, Aug. (2002).

[6] C.E. Perkins, E.M. Royer, "Ad-hoc on-demand distance vector routing, in", Proceedings of the 2nd IEEE Workshop on Mobile Computing Systems and Applications, vol. 2, February, (1999), pp. 90–100, New Orleans, LA.

[7] D.B. Johnson, D.A. Maltz, "Dynamic source routing in ad hoc wireless networks", Kluwer International Series in Engineering and Computer Science (1996) , pp. 153–179.

[8] Uri Yechiali, "A Queuing-Type Birth-and-Death Process Defined on a Continuous-Time Markov Chain", Queueing Systems - Theory and Applications - QUESTA , vol. 60, no. 3-4, pp. 271-288, (2008).

[9] D. Cavendish and M. Gerla, "Internet QOS Routing using the BELLMAN-FORD ALGORITHM", Proceedings IFIP Conference on High Performance Networking, (1998), pp. 1-5, Vienna, Austria.

[10] R. Beraldi, "On message delivery through approximate information in highly dynamic mobile ad hoc networks", The Seventh International Symposium on Wireless Personal Multimedia Communications, Italy, vol. 12, (2004), pp. 15.

[11] L. Rosati, M. Berioli, G. Reali, "On ant routing algorithms in ad hoc networks with critical connectivity", Ad Hoc Networks 6 (6) (2008) ,pp. 827–859.

[12] I. Bouazizi, "ARA-the ant-colony based routing algorithm for MANETs", Proceedings of the 2002 International Conference on Parallel Processing Workshops, IEEE Computer Society Washington, DC, USA, (2002).

[13] H.F. Weddle, M. Farooq, T. Pannenbaecker, B. Vogel, C. Mueller, J. Meth, R. Jeruschkat, Bee AdHoc, "An energy efficient routing algorithm for mobile ad hoc networks inspired by bee behavior", Proceedings of the 2005 Conference on Genetic and Evolutionary Computation, ACM New York, NY, USA, (2005), pp. 153–160.

[14] G. Di Caro, F. Ducatelle, L.M. Gambardella, AntHocNet,"An adaptive nature-inspired algorithm for routing in mobile ad hoc networks", European Transactions on Telecommunications 16 (5) (2005) , pp. 443.

[15] M. Roth, S. Wicker, "Performance evaluation of pheromone updates in swarm intelligent MANETs", The Sixth IFIP IEEE International Conference on Mobile and Wireless Communication Network, (2005), Atlantic City, USA.

[16] M. Roth, S. Wicker, Termite, "Ad-hoc networking with stigmergy", Global Telecommunications Conference, 2003.GLOBECOM'03.IEEE, vol. 5, (2003), San Francisco, USA.

[17] L.Reddeppa Reddy, S.V. Raghavan, "SMORT: scalable multipath on demand routing for mobile ad hoc networks", Ad Hoc Networks 5 (2) (2007), pp. 162–188.

[18] S.Y. Wang, C.L. Chou, C.H. Huang, C.C. Hwang, Z.M. Yang, C.C. Chiou, and C.C. Lin , "The Design and Implementation of the NCTUNs 1.0 Network Simulator", Computer Networks, Vol. 42, Issue 2, (June 2003), pp. 175-197 (SCI).

[19] S.Y. Wang, C.L. Chou, C.H. Huang, C.C. Hwang, Z.M. Yang, C.C. Chiou, and C.C. Lin ,NCTUns 6.0, "A Simulator for Advanced Wireless Vehicular Network Research", IEEE WiVEC 2010 (International Symposium on Wireless Vehicular Communications), (May 16-17, 2010), Taiwan, Taiwan. (Invited demo paper).

[20] J.H Tarng, B.W Chuang, F.J Wu, "A Novel Stability-Based Routing Protocol for Mobile Ad-Hoc Networks" ,IEICE Transactions on Communications, (2007), E90-B(4): pp. 876-884.

[21] E.D.KanmaniRuby, S.Rajasurya, K.Swarnam, "A route stability-based optimized AODV protocol", Proceedings RTCSP'09, Amrita Vishwa Vidyapeetham, India , pp. 1-5(2009).

[22] N. Wang and J. Chen, "A Stable On-Demand Routing Protocol for Mobile Ad Hoc Networks with Weight-Based Strategy", IEEEPDCAT'06, pp. 166–169, (Dec. 2006).

[23] Tseng Y-C, Ni S-Y, Shih E-Y,"Adaptive approaches to relieving broadcast vol.storm in a wireless multihop mobile ad hoc network".,IEEE Trans Comput, pp. 299–31, (2003).

[24] Castañeda R, Das SR, "Query localization techniques for on-demand routing protocols in ad hoc networks",Proceedings of the 5th annual ACM/IEEE international conference on mobile computing and networking, pp 186–194, (1999), New York, NY, USA.

[25] S.J. Lee and M. Gerla, "AODV-BR : Backup Routing in Ad hoc Networks", In Proceedings of the IEEE Wireless Communications and Networking Conference, pp. 1311- 1316, (2000), Chicago, IL , USA.

[26] LIU Jian and LI Fang-min, "An Improvement of AODV Protocol Based on Reliable Delivery in Mobile Ad hoc Networks", Fifth International Conference on Information Assurance and Security, pp. 507-510, (2009), China.

[27] Qing Li, Cong Liu, Han-Hong Jiang, "The Routing Protocol of AODV Based on Link Failure Prediction [C]", Proceeding of the International Conference on Software Process, (2008), Leipzig, Germany.

[28] S. Sethi and S. K. Udgata, " Optimized and Reliable AODV for MANET", International Journal of Computer Application, 3, No. 10, (July 2010).

[29] Muneer Bani Yassein, Sanabel Nimer, and Ahmad AL-Dubai, "The Effects of Network Density of a New Counter-based Broadcasting Scheme in Mobile Ad Hoc Networks", Proceeding of the 10th IEEE International Conference on Computer and Information Technology (CIT2010), (29 June – 01 July, 2010), Bradford, UK.

[30] Perkins, C. E. , Royer, E. M. and Das, "Ad Hoc On-Demand Distance Vector Routing. IETF Internet Draft.,http://www.ietf.org/internet-drafts/draft-ietf-manetaodv-03.txt, (1999).

[31] Toh, C. K., " Associativity-based routing for ad hoc mobile networks", Wireless Personal Communications
Journal, Special Issue on Mobile Networking & Computing Systems, 4, 2 (March 1997), pp. 103–109.

[32] Spohn, M. and Garcia-Luna-Aceves, " Neighborhood aware source routing", in Proceedings of 2nd ACM International Symposium on Mobile Ad Hoc Networking and Computing (MobiHoc),(2001), Chicago, IL, USA.

[33] Muneer Bani Yassein, Mustafa Bani Khalaf , Ahmed Y Al-Dubai, "A new probabilistic broadcasting scheme for mobile ad hoc on-demand distance vector (AODV) routed networks", The Journal of Supercomputing (2010) Volume: 53, Issue: 1, pp. 196-211.

[34] Ni, S.-Y., Tseng, Y.-C., Hen, Y. S. V , & Sheu, J.-P. "The broadcast storm problem in mobile ad-hoc networks", In Proceedings of MobiCom, (2001), Rome, Italy.

[35] Spohn , M. A. "Domination in graphs in the context of mobile ad hoc networks", Ph.D. dissertation, University of California, Santa Cruz (2005).

[36] Karp, Kung, "Greedy perimeter stateless routing for wireless networks" , In Proceedings of the sixth annual ACM/IEEE international conference on mobile computing and networking, pp. 243–254, August(2000), Boston, MA, USA.

11

An Efficient Approach for Data Aggregation Routing using Survival Analysis in Wireless Sensor Networks

Dr.B.Vinayaga Sundaram, Rajesh G, Khaja Muhaiyadeen A, Hari Narayanan R, Shelton Paul Infant C,Sahiti G, Malathi R, Mary Priyanga S

Department of Information Technology, Anna University, Chennai
bvsundaram@annauniv.edu, raajiimegce@gmail.com, khaja.it@gmail.com, hari.zlatan@gmail.com, sheltonpaul89@gmail.com, sahiti.mit@gmail.com, angelmalathi@gmail.com, marypriyanga@yahoo.com

ABSTRACT

Wireless Sensor Network (WSN) is a collection of small sensor nodes with a communications infrastructure to achieve mutual communication and to monitor and record conditions at diverse locations. The major constraints of WSN are limited availability of power and it is prone to frequent node failures. In order to prolong the lifetime of the sensor nodes, the sensor data should efficiently reach the base station and there should be a reduction in message transmission, which consumes the majority of the battery power. Aggregation of data at intermediate sensor nodes helps in saving the energy that would be spent if the nodes send directly to the base station. In addition to aggregation, the mechanism to overcome node failures is also essential to ensure the successful delivery of the data packets to the base station. This paper proposes an efficient way based on inverse-square law along with survival analysis for aggregating data without the formation of an explicit structure and to overcome node failures. Our evaluation of performance shows a considerable decrease in the number of transmissions required to carry the sensed data to the base station and also a considerable increase in packet delivery ratio. By using this approach, it is possible to minimize power usage of sensor nodes effectively.

KEYWORDS

Inverse Square law, Survival Analysis, Kaplan-Meier estimator, Enhanced Random Delay, Wireless Sensor Networks, Node failure

1. INTRODUCTION

Wireless Sensor Network (WSN) consists of spatially distributed autonomous sensors to cooperatively monitor physical or environmental conditions, such as temperature, sound, vibration, pressure, motion or pollutants. A number of sensor nodes are densely deployed in a field of interest and they observe the phenomena at different points in the field, which are sent to a data sink or base station, located either at the centre or out of the field, for processing [8]. Wireless sensor networks are now used in many civilian application areas, including environment and habitat monitoring, health applications, home automation, and traffic control. Size and cost constraints on sensor nodes result in corresponding constraints on resources such as energy, memory, computational speed and bandwidth. Specific applications for WSNs include habitat monitoring, object tracking, nuclear reactor control, fire detection, and traffic monitoring. In a typical application, a WSN is scattered in a region where it is meant to collect data through its sensor nodes. Instead of directly sending the data to the sink, it is highly desirable to aggregate the data through effective data-aggregation techniques to minimize the power consumption of the sensor nodes during data transmission and thereby prolonging the overall network lifetime. Since the sensor nodes cannot be recharged, the sensed data should reach the base station through multihop routing. Many approaches were common. Mostly, the sensor network is grouped into clusters or a tree like structure and data moves up hierarchically to the base station. Here those nodes in top of the hierarchy, gets affected more. In order to handle this, we propose the structure-less approach where data gets aggregated quickly. A two step framework is used. In first, node that is more probable to contain data is found and then the time for which a node must wait for data to arrive from other nodes is found. We show the gain in performance by means of the reduction in number of transmissions in the network. Finally, our simulation results will also substantiate our claim of the gain in performance. In addition, sensor nodes in WSNs are prone to failure due to energy depletion, hardware failure, software bugs, communication link errors, environmental interference, malicious attack, and so on. Fault tolerance is the ability of a system to deliver a desired level of functionality in the presence of faults. Since the sensor nodes are prone to failure, fault tolerance should be seriously considered in the sensor network applications. Hence a fault tolerant mechanism is proposed to facilitate the transmission of the data packets to the base station so that the data packets reach the destination without any loss even in the presence of intermediate node failures along the multihop route to the base station.

2. RELATED WORK

All of the previous work done can be broadly classified into 2 categories viz structured approach and structure less approach. This section delves deeper into the various approaches proposed in these categories.

2.1 Structured Approach:

Rumor routing [2] routes the queries to nodes that observed a particular event rather than flooding, but maintaining agents and event-tables are sometimes infeasible. In [3], a protocol called *Low Energy Adaptive Clustering Hierarchy (LEACH)* forms cluster and randomly selects cluster heads and rotates this role to evenly distribute the load among nodes. In [4], *Power-Efficient Gathering in Sensor Information Systems* enhancement of *LEACH,* where nodes communicate only with their closest neighbors and once the turns of all the nodes are over, a new round will start. All the structured approaches results in fixed delay, which would be intolerable in large network deployments.

2.2 Structure-less Approach:

Studying the effect of the changing network topology is essential for analyzing the performance of structure less algorithms. [5] proposes a *Distributed Random Grouping(DRG)* algorithm that uses a probabilistic grouping to answer aggregate queries like computation of sum, average, maximum, minimum, etc. Through randomization, all values will progressively converge to the correct aggregate value (the average, maximum, minimum, etc.) in this method. The disadvantage with this approach is that it involves periodic and frequent transfer of message exchange between the nodes of a group. [6] proposed a novel data aggregation protocol for event based applications via 2 mechanisms namely *Data Aware Anycast (DAA)* at the MAC layer and *Randomized Waiting (RW)* at the application layer. DAA mechanism used RTS and CTS packet transmissions in order to determine whether the neighbor node has data. Since sensor nodes need to wait for data from other nodes, RW was proposed where each node chooses a random delay value within a maximum delay τ. We call this as *Structure Free Data Aggregation (SFDA)* and compare our approach with SFDA to show the improvement in performance.

3. PROBLEM FORMULATION

Since we are primarily concerned with a structure less network, our aim here is to find a neighbor node within its communication range which is most probable to contain data and given that a node has sensed data, it needs to know how long it should wait for data from other nodes.

In addition to finding the next node to forward the data, another major concern of wireless sensor networks is to detect node failures so that the packets reach the base station without any loss. Our aim is to detect node failure with less number of control message transmissions and increase packet delivery ratio.

3.1 Neighbor Node Detection

In order to detect which node is most probable to contain data, we propose a prediction based approach as described in [1]. [6] proposed DAA method for the same problem of finding the neighbor node with data. The main problem with this approach is that it uses RTS and CTS packet transmissions for every data transmission. This can induce a serious load and can reduce the lifetime of sensor nodes considerably. Hence, we propose an approach which does not involve any kind of communication between the nodes. Each node analyses the collected upstream nodes data and calculates a probability value and using this probability value, it decides the node for which it has to forward data.

Inverse-Square Law: In physics, an Inverse-Square Law is any physical law stating that some physical quantity or strength is inversely proportional to the square of the distance from the source of that physical quantity. This law is very suitable for our environment since all electromagnetic waves has to obey the inverse square law. Since we do not know the position of the source, we take the position of node with highest sensed value as the position of source, as the node is closest to the source. The value sensed by the sensor node which is at a particular distance from the event is analogous to the intensity value at that distance from the source. Next, we need to determine the rate at which the sensed value changes with respect to the distance. If we obtain this, then we can predict the most probable sensed value for the downstream nodes. The Inverse-Square Law is,

$$I = \frac{k}{r^2} \tag{1}$$

Where I is Intensity, r is distance from source and k is a constant. Differentiating (1) gives the rate at which the sensed value changes with respect to distance.

$$\frac{dI}{dr} = -2 K_m r^{-3} \tag{2}$$

The negative sign indicates that the intensity decreases with respect to distance. Here K_m represents the mean value of K for all the upstream nodes.

$$K_m = \frac{\sum_{j=1}^{N} I_j r_j^2}{N} \tag{3}$$

From this, we can determine the rate at which the intensity has varied with respect to distance for all the upstream nodes. It is most likely to vary at the same rate for downstream nodes also.

Survival Analysis (SA) It is a branch of statistics which deals with death in biological organisms and failure in mechanical systems with respect to time t. We correlate this to our environment where we define the survival function S(r) as the probability that data is sensed by a node which is beyond distance r. It is defined as,

$$S(r) = P(R > r) \tag{4}$$

Similar to the lifetime distribution function of SA, we define the *Intensity distribution function*, which is a compliment of the survival function S(r), as

$$F(r) = P(R \leq r) = 1 - S(r) \tag{5}$$

The above equation (5) indicates the probability that data is sensed by a sensor node that is located at distance r or below. *Event density function* which denotes the rate of data sensed with respect to distance is given by

$$f(r) = F'(r) = \frac{d}{dr} F(r) = 2K_m r^{-3} \tag{6}$$

Kaplan Meier Estimator (KPE) is a probabilistic measure of SA which denotes the probability that a living being survives up to some point of time. In other words, it denotes the probability that $P(S \leq t)$. For our environment,
According to KPE,

$$S^\wedge(r_i) = \prod_{x \leq ri} (1 - m^\wedge(x)) \tag{7}$$

Here, Hazard function $m^\wedge(x)$ is defined as the event rate at time *t* conditional on survival until time *t* or later (i.e., $T \geq t$). For our environment is given by,

$$m^\wedge(x) = \frac{d(x)}{n(x)} \tag{8}$$

Where d(x) is number of nodes without event detection and n(x) is number of events under study. We now calculate the probability that data exists in the next downstream node. The probability that data don't exists in next downstream node is,

$$P(R \leq r_0 + r/R > r_0) = \frac{\int_{r_0}^{r_0+r} f(r)dr}{S(r_0)} \tag{9}$$

The node with the least probability is chosen as the node that is most probable to have data. We then send data to that node and data aggregation is most probably achieved.

3.2 Delay Calculation

Once a node has collected data, it needs to know how long it should wait for data to come from its downstream nodes. In SFDA, they used a randomized waiting scheme wherein each node takes a random delay value within a certain maximum delay value. Deterministically assigning the waiting time to nodes such that nodes closer to the sink wait longer can avoid the problem but results in a fixed delay for all packets, which would be intolerable in large network deployments. Therefore, randomized waiting scheme is the optimal approach for assigning delay values to the sensor nodes. However, we propose Enhanced Random Delay (ERD), a subtle difference to that approach wherein, instead of making the maximum delay value fixed for the entire network, we make the maximum delay value dependent on the distance of the node from the sink. This provides an improvement in performance because of lesser probability for a node to choose a delay value that will make it wait longer than is necessary.

3.3 Node failure mechanism

Nodes in WSNs are prone to failure due to energy depletion, hardware failure, software bugs, communication link errors, environmental interference, malicious attack, and so on. Fault tolerance is the ability of a system to deliver a desired level of functionality in the presence of faults. Since the sensor nodes are prone to failure, fault tolerance should be seriously considered in the sensor network applications. Hence a fault tolerant mechanism (FTM) to facilitate the transmission of data packets to the base station is proposed so that the data packets reach the destination without any loss even in the presence of intermediate node failures along the multihop route to the base station.

Since sensor nodes are frequently subjected to unexpected failures, the wireless sensor network should be able to function and transmit messages within the network in spite of node failures. Under this scheme it is assumed that the failed nodes fall under either of the following two categories:

- The failed node is an intermediate node alone that just forwards the messages from the sender node to the destination which in most of the cases will be the base station
- The failed node is the one that senses the data along with performing aggregation and forwarding.

Under the second case, the failed sensor node cannot be used as an aggregation point and these nodes can be detected as failed only if this node falls under the route of any other sensor node that transmits. In case of node failures, we provide a solution to detect the failed node with comparatively less overhead of message transmissions and transmit the data packets successfully to the base station without affecting its transmission because of the node failure.

The proposed fault tolerant mechanism (FTM) is that whenever a node calculates a probability value to identify the most probable downstream node to which it can transfer the data, the sender node will check the presence of that node by sending a beacon signal and starting a timer. If the node doesn't respond to the beacon signal within the stipulated time then, the node is considered to be failed and thus, the sender will select another node in an alternate route to transfer the data packets. While selecting the new destination node to which it intends to send the data, the sender will select the node based on the same criteria of calculating the probability based on equation (9) that the selected node will contain data. Additionally, when a node detects that a particular node has failed it sends message to its nearby nodes about the id of the failed node so that this failed node id can be removed from their routing table list. This transmission of messages to indicate the node failure is in a localized area and thus will not increase the overall overhead of the entire network to a great extent. By this localized transmission, it further avoids the transmission of beacon signals by other nodes to detect that failed node again and again every time they try to transmit to this failed node. Since a nearly equal alternate route is found, the packet delivery ratio increases in spite of node failures.

4. PERFORMANCE ANALYSIS

In order to evaluate the performance of our approach, we cannot use the number of aggregation points as the metric because the packets may be aggregated after travelling many hops. Expected number of transmissions and packet delivery ratio is the correct metric for evaluating the performance of our algorithm. To evaluate the performance of detecting the node failures, packet delivery ratio is to be used to compare if most of the packets that are sent by the event sensing nodes, reach the destination.

4.1 Expected Number of Transmissions

In this section, we will first calculate the probability for the packet to get aggregated at a node. After this, we will calculate the expected number of transmissions. In SFDA [5], they assume that the delay chosen by each node is distinct from each other. Using this assumption, they calculate the expected number of transmissions in the network. In a practical situation, each node is independent of each other i.e., each node chooses a random number that is independent of the random value chosen by another node. So, in the worst case, we compare the expected number of transmissions in the network of our approach with that of theirs.

Consider a chain topology of nodes from v_0 to v_n where v_0 is the sink and all nodes have data to send. Let the number of nodes in the network be 7. Let Y be the discrete random variable representing the number of hops a packet has been forwarded before it is aggregated. As an example, for 7 nodes shown, the node v_n can choose its delay so that its sending order (l) ranges from 1 to 6, and the node v_{n-h} can take its order (k) from 2 to 7. The remaining (h-1) nodes in between take delay values in l^{h-1} ways. The number of ways N_0 in which the nodes from v_n to v_{n-h} take their delay values is therefore,

$$N_0 = \sum_{k=2}^{n} \sum_{l=1}^{k-1} l^{h-1} , \; if \; 0<h<n \tag{10}$$

From [1] it can be shown that the using equation (10) the expected number of transmissions is derived as,

$$E_0 = \sum_{h=1}^{n-1} h \times \left[\frac{N_0}{n^h} \right] + \frac{\sum_{h=1}^{n} h^{n-1}}{n^{n-2}} \tag{11}$$

Based on equation (11) as derived in [1], the maximum allowed delay value for all nodes is chosen as τ. In our approach, we fix the maximum delay for each node based on the node's distance from the sink. The node will choose a random delay within that fixed maximum delay. Obviously, since we reduce the limit of delay value for each node, nodes farther from the sink will choose a lower delay than the nodes that are closer to the sink thereby attaining early aggregation. Therefore as described in [1], the expected number of transmissions for our approach is given by,

$$Ep = \sum_{i=2}^{R} \left[\sum_{h=1}^{R-1} \frac{h(i-1)!}{(i+h)!} \times N_p \right] + \sum_{h=1}^{R} \frac{h^2}{(h+1)!} \tag{12}$$

Where Np denotes the number of ways in which the nodes from v_n to v_{n-h} takes its order whose value is shown in [1]. Thus for increasing network size the expected number of transmissions in the ERD decreases drastically compared to the SFDA approach.

4.2 Packet delivery ratio (PDR)

Packet delivery ratio can be defined as the ratio of the number of sensed data packets received by the base station to the number of data packets sent by the event sensing nodes.

$$PDR = \frac{P_{recv}}{P_{sent}} \tag{13}$$

PDR is in the range of 0 to 1. The higher the value of PDR indicates the packets are delivered successfully to the base station through different routes that doesn't include the failed nodes. Therefore, this parameter is compared with the number of nodes that are failed in the network.

5. PERFORMANCE EVALUATION

The OMNeT++ 4.0 simulator is used along with MiXiM simulation framework to provide mobility among the events. The nodes are arranged in a grid topology with inter-node separation of 30 m between the sensor nodes. Intel-Lab data [7] is also used to calculate the number of transmissions for this network using our approach. The simulation scenario used is same as described in [1].

5.1 Simulation Scenario

Table 1. Default Parameters Used in the Simulation

Parameters	Values
Network Topology	300 m x 300 m
Data Rate	38.4 Kbps
Communication Range	55 m
Mobility Model	ConstSpeedMobility
Packet Size	50 bytes
Sensing Interval	10 s
Event Size	50 m to 200 m
Internode Separation	30 m
Event Moving Speed	10 m/s
Maximum Delay	0.8 s to 4 s

The Packet Aggregation Ratio (PAR) is used as metric to compare different protocols. PAR determines how effective a protocol is in aggregating packets and is (Number of nodes in which a packet gets aggregated/Number of nodes through which packet is transmitted to sink). PAR will be in the range 0 to 1.Maximum the value of ratio determines the packet is effectively aggregated in the route. Table.1 shows the various in the default parameters used in the simulation.

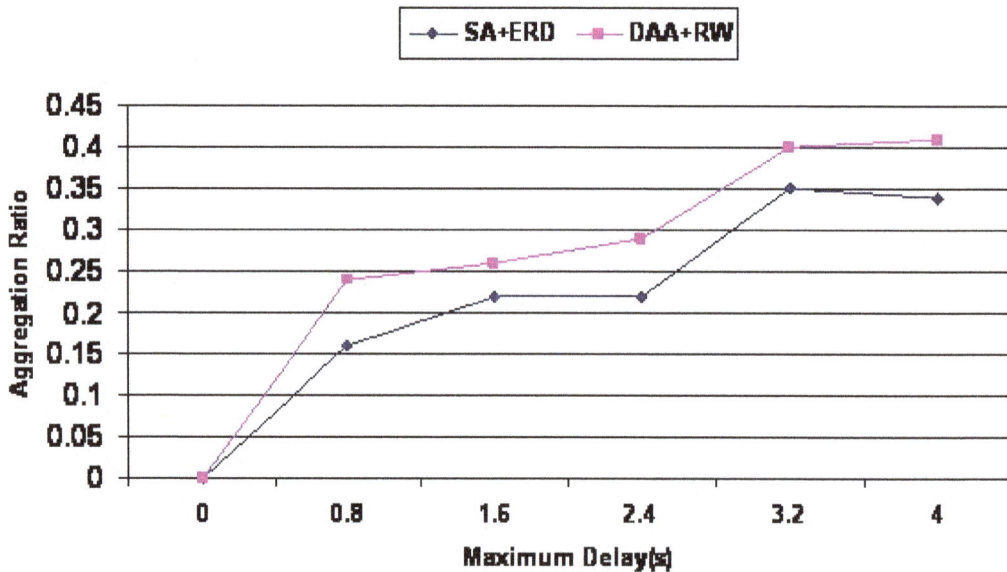

Figure1. Aggregation ratio vs Different Maximum Waiting time

The Packet Delivery Ratio (PDR) is used as a metric to compare the proposed node failure avoidance mechanism with the network without using this node failure mechanism. The PDR value is calculated for different number of nodes that have failed in the network. The simulation results show that the packet delivery ratio is high when the node failure avoidance mechanism is used.

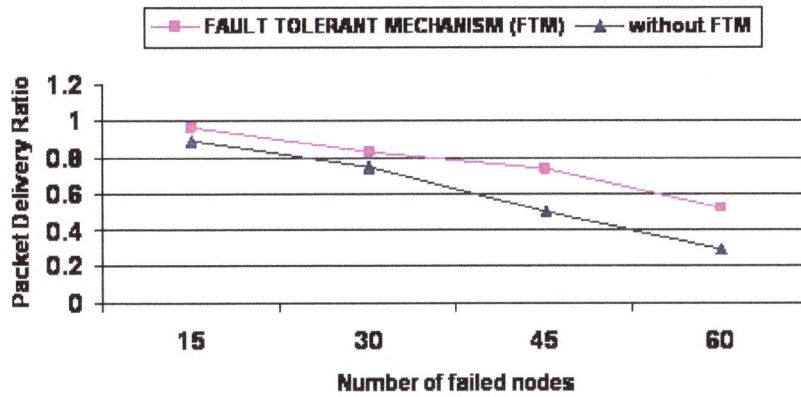

Figure 2. Packet delivery ratio vs number of failed nodes

Maximum Delay (MD) is used as a varying parameter and the corresponding number of transmissions for the Intel-Lab network is plotted as shown in Figure 3. For increase in MD, the total number of transmissions in the network should decrease. From the graph Figure 3, it is shown that, the proposed approach performs better than SFDA approach. On average, the number of transmissions decreases by 9.1 percent.

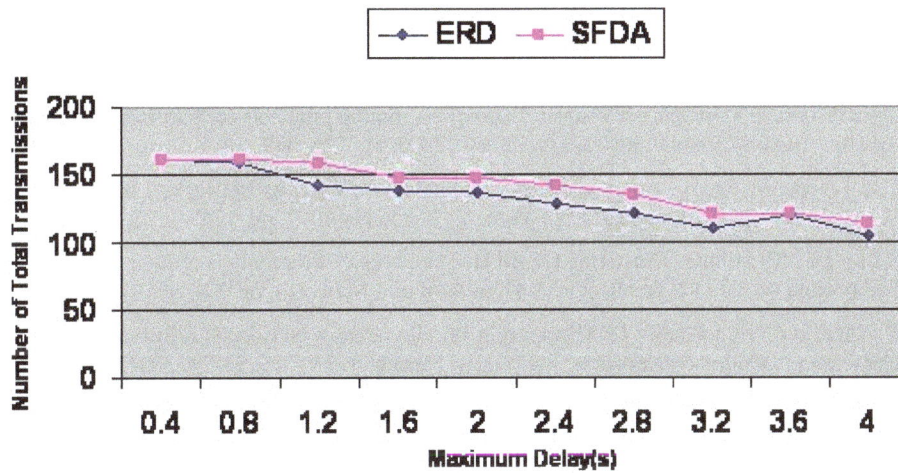

Figure 3. Number of Transmissions for Varying Maximum Delays

6. CONCLUSION

In this paper, we proposed a cost forming any explicit structure in the network. Instead of spending more energy in communication, SA, neighbor node that is more probable to contain data. Using the ERD method, the unwanted message transmissions are further reduced by finding the random delay within a fixed delay. From extensive simulation exhibits almost equal performance to SFDA by minimizing energy consumption and unnecessary message transmissions. By the node failure avoidance mechanism we have proposed, it results in high packet delivery ratio in spite of node failures.

REFERENCES

[1]   Khaja Muhaiyadeen A, Hari Narayanan R, Shelton Paul Infant C, Rajesh G, Inverse Square Law Based Solution for Data Aggregation Routing using Survival Analysis in Wireless Sensor

Networks, International Conference on Networks and Communication (Netcom), Bangalore, India,2010

[2]   Braginsky, D., Estrin, D.: Rumor Routing Algorithm for Sensor Networks. In: Proceedings of the First Workshop on Sensor Networks and Applications (WSNA), Atlanta, GA, October 2002.

[3]   Heinzelman, W., Chandrakasan, A., Balakrishnan, H.: Energy Efficient Communication Protocol for Wireless Microsensor Networks. In: Proceedings of the 33rd Hawaii International Conference on System Sciences (HICSS'00), January 2000.

[4]   Lindsey, S., Raghavendra, C.: PEGASIS: Power-Efficient Gathering in Sensor Information Systems, IEEE Aerospace Conference Proceedings, 2002, Vol.3, 9-16 pp.1125-1130.

[5]   Jen-Yeu Chen., Jianghai Hu.: Analysis of Distributed Random Grouping for Aggregate Computation on Wireless Sensor Networks with Randomly Changing Graphs. In: IEEE TRANSACTIONS ON PARALLEL AND DISTRIBUTED SYSTEMS, VOL. 19, NO. 8, AUGUST 2008.

[6]   Kai-Wei Fan., Sha Liu., Prasun Sinha.: Structure-Free Data Aggregation in Sensor Networks, In: IEEE TRANSACTIONS ON MOBILE COMPUTING, VOL. 6, NO. 8, AUGUST 2007.

[7]   Intel Berkeley Research Lab Sensor data, http://db.csail.mit.edu/labdata/labdata.html.

[8]   F. Akyildiz, W. Su, Y. Sankarasubramaniam, and E. Cayirci, ''Wireless sensor networks: A survey,'' Computer Networks Journal, volume 4, no. 12, Mar. 2002, pp. 393—422.

[9]   Giuseppe Anastasi a, Marco Conti, Mario Di Francesco, Andrea Passarella, Energy conservation in wireless sensor networks: A survey, Adhoc networks journal Elsevier, volume 7, 2009 , pp. 537–568.

[10]   Kemal Akkaya *, Mohamed Younis, A survey on routing protocols for wireless sensor networks, Ad Hoc Networks journal, Elsevier, volume 3, 2005, pp. 325–349.

[11]   C.R. Lin and M. Gerla, "Adaptive Clustering for Mobile Wireless Networks", IEEE Journal on Selected Areas In Communications, volume 7, September 1997, pp. 1265-1275.

[12]   S.Basagni, "Distributed Clustering for Ad Hoc Networks", Proceedings of International Symposium on Parallel Architectures, Algorithms and Networks, pp. 310-315, June 1999.

[13]   D. Chen, and P. Varshney, "QoS Support in Wireless Sensor Networks: A Survey," International Conference on Wireless Networks, Las Vegas, Nevada, USA, June 21-24, 2004.

[14]   G. Bravos, A. Kanatas, "Energy Consumption and Trade-offs on Wireless Sensor Networks," IEEE 16th International Symposium on Personal, Indoor and Mobile Radio Communications, Vol. 2, pp. 1279- 1283, September 2005.

[15]   Ameer Ahmed Abbasi a,*, Mohamed Younis, A survey on clustering algorithms for wireless sensor networks, Computer Communications , Science Direct, Volume 30 2007 , pp.2826–2841.

12

ROUTING TECHNIQUES IN COGNITIVE RADIO NETWORKS: A SURVEY

Amjad Ali[1]Muddesar Iqbal[2]Adeel Baig[1]Xingheng Wang[3]

[1]School of Electrical Engineering and Computer Sciences National University of Science and Technology, Pakistan
Email: {amjad.ali,adeel.baig}@seecs.edu.pk

[2]Faculty of Computer Science & Information Technology, University of Gujrat, Pakistan

[3]College of Engineering, Swansea University, Swansea, UK
Email: xingheng.wang@swansea.ac.uk

ABSTRACT

Cognitive Radio Networks (CRNs) are being studied intensively. The major motivation for this is the heavily underutilized frequency spectrum. CRN has the capability to utilize the unutilized frequency spectrum. Routing in CRN is a challenging task due the diversity in the available channels and data rates. In this paper, we present a survey of the state-of-the-art routing techniques in CRNs. We first outline the design challenges for routing protocols in CRNs followed by a comprehensive survey of different routing techniques. Furthermore we classified these routing protocols into spectrum aware-based, multipath-based, local coordination-based, reactive source-based and tree-based routing techniques depending on the protocol operation.

KEYWORDS

CRNs, DSA, RF spectrum, Routing

1. INTRODUCTION

Cognitive Radio Networks (CRNs) can operate in the licensed frequency band to improve its utilization with the coexistence of the Primary Users (PRs) or licensed users. PRs have the main rights over the licensed band in which they are operating.

Radio frequency (RF) is an important resource that people uses all around the world for many services i.e. safety, communication, employment, and entertainment [1].The dedicated frequency band is allocated to the paid user that uses this frequency for specific service. Thus the RF band allocated can be vastly underutilized. Recent studies show that only 5% of the spectrum from 30 MHz to 30 GHz is used in the US [2]. The Federal Communications Commission (FCC) of United States of America found that spectrum usage is a more significant problem than the actual physical availability of RF spectrum [3]. The spectrum availability problem arises due to the currently deployed static spectrum allocation policy that limits the usage of the licensed RF band only to the licensed user or primary user. These findings need more efficient methods for utilization of the RF resources and the Cognitive Radio (CR) technology is envisioned as new mechanism for flexible usage of the RF spectrum. This technology enables the secondary users or unlicensed users to operate in the licensed band with the coexistence of the licensed users or primary users. Secondary users have the ability to identify and utilize the available channels in the RF spectrum. The ability of a secondary user to change its frequency of operation is commonly referred as dynamic spectrum access (DSA). Thus we can say that cognitive network is a network that can observe current network conditions, and then act on those conditions. The network can learn from these actions and use them to make future decisions [4]. Ryan W. Thomas and Daniel H, define the CRNs in the context of machine learning as "cognitive network improves its performance through experience gained over a period of time without complete information about the environment in which it operates" [5]. Thus a secondary user can change its transmitter parameters based on its learning from the environment, based on these changes the secondary user can efficiently utilize the frequency band and avoid from the interference on the primary user.

The main feature of CR technology that how it operates in the licensed band with the coexistence of the PR users is that it identifies the spectrum holes in the RF spectrum called white spaces. These white spaces are the wastage in the RF spectrum and would be used by secondary user for its communication.

Figure 1shows the concept of spectrum holes.

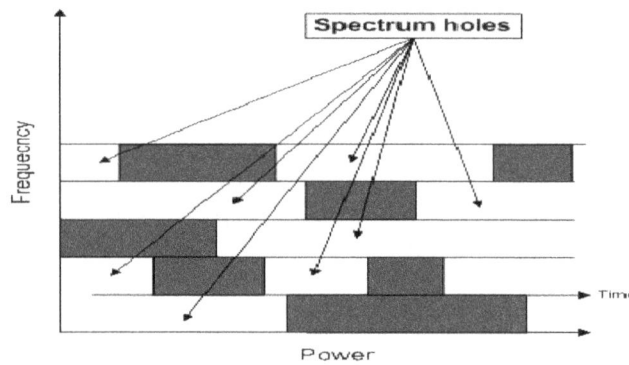

Figure 1: Spectrum holes concept

The rest of this paper is organized as follows. In Section 2, we discuss the CRNs architecture. In section 3, we discuss routing differences and challenges. A comprehensive survey of routing techniques in CRNs is presented in Section 4.

2. ARCHITECTURE FOR CRNs

A clear description of the CRNs architecture is important for the understanding of their working and designing novel protocol for communications. Figure2 describes the architecture of the CRNs.

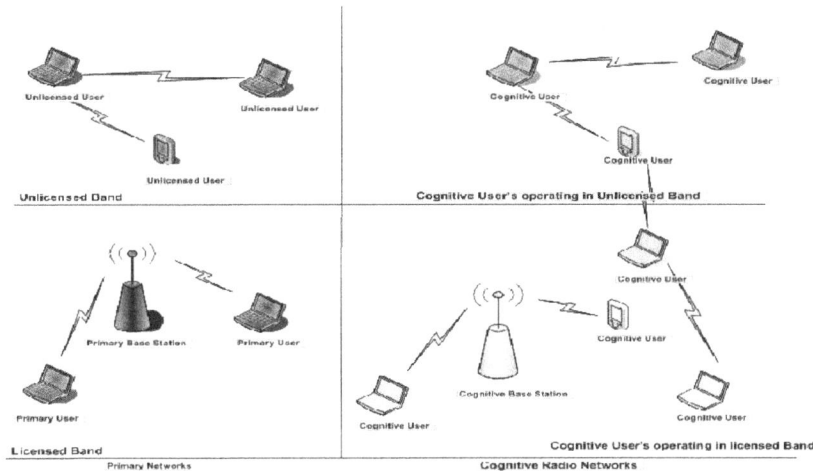

Figure 2: Cognitive network architecture.

The CRNs can be classified into two main categories:

    a.   Infrastructural CRN

    b.   Infrastructure less CRN

Infrastructural CRN: In this type of CRN the communication is done with the help of a fixed infrastructure component called CRN base station.

Infrastructure less CRN: In this type of CRN the communication is done without the support of fixed infrastructure or base station. This type is similar to ad hoc multi-hop network.

The elements of the primary and the cognitive network are defined as follows:

Primary Network: A network that owns a separate RF spectrum band for its services (e.g. common cellular and TV broadcast networks).The components of the primary network are as follows:

Primary User: Primary user or licensed user has main rights to operate in its RF spectrum band. The primary user cannot bear any interference on its RF band due to any secondary user.

Primary Base Station: It is a fixed infrastructure component that controls the access of its licensed users. This component does not have any ability to share RF spectrum band with CR users.

Cognitive Network: Cognitive network does not have any RF spectrum band for their communication. That's why these networks are called unlicensed networks.

The components of cognitive networks are as follows:

Cognitive User: Cognitive user does not have its own frequency band; therefore these users are called unlicensed users or secondary users. These users only share the licensed band in an opportunistic manner. Secondary user requires extra functionality such as spectrum sensing to operate in licensed band.

Cognitive Base Station: Cognitive base station is a fixed infrastructure component similar to primary base station with cognitive capabilities. The secondary users access other networks through it. It provides single hope connection to secondary users.

3. ROUTING DIFFERENCES AND CHALLENGES IN CRNS

Conventional infrastructure networks and mobile ad hoc networks have been studied since last decade and many routing protocols, e.g. proactive, reactive, hierarchical and multicast are available for such networks [6].

Now it's time to investigate the novel routing protocols for CRNs, as routing is a challenging task in such networks, particularly in multi hop CRNs due to the diversity in channel availability and data rates. Multi-interface enabled CR user can avail multiple available channels simultaneously, thus increase overall network performance and reduce the interference on the primary users. Due to this vital feature of CRNs conventional routing metrics such as hope count, congestion, etc, are not sufficient for routing decision in CRNs [6].

The primary differences and challenges between routing of CRNs and routing of conventional wireless networks can be summarized as follows:

3.1 Link Availability:

DSA senses the RF band and fetches the available opportunities. Thus these networks use the licensed band in opportunistic manner for communication. These communication opportunities are available when primary users have fewer usage of their frequency band or they are not using it at all, therefore availability of channel is time and geographic based. Due to this random availability of channel the CRNs topology is random even all communicating nodes are static [7].

3.2 Deafness Problem:

Due to the diversity in channel or link availability, links may be available only a fraction of time in CRNs. This random availability may cause the deafness problem, which is switching among available channel set to maintain the route or avoid the interference on primary users. This causes extra delay in CRNs communications [8].

3.3 Unidirectional Links:

In CRNs, unidirectional links are more likely to be happened. In CRNs links are only available for a fraction of time instead of minutes or hours. This cause's unidirectional links as there is no guarantee that a channel used for sending station will be available till the receiving station uses the same channel for transmission.

3.4 Heterogeneous Wireless Networks:

CRNs are usually come into existence by multiple and heterogeneous wireless networks. Therefore intersystem handover is critical and required for routing. Along with this channels or links are available for an extremely short duration; therefore successful networking lies in cooperative relaying among such heterogeneous wireless networks [9].

3.4.1 Security issues:

Enabling of CRNs at the price of losing security is a debatable issue. Short duration of available channel is not a sufficient for CR user or node to get secure certificate.

4. ROUTING PROTOCOLS FOR CRNS

In this section, we survey the state-of-the-art routing protocols for CRNs. We classified these routing protocols into spectrum aware-based, multipath-based, local coordination-based, reactive source-based and tree-based routing techniques depending on the protocol operation as shown in figure3. Furthermore we presented protocol summary in the form of tables.

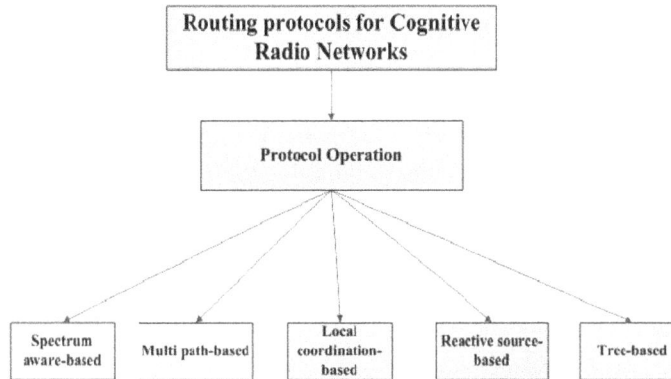

Figure 3: Routing protocol for CRNs

4.1 Dynamic Spectrum-aware Routing

Dynamic spectrum-aware routing protocols enable CR technologies to effectively utilize unallocated wireless spectrum. In such routing protocols route discovery is incorporated with spectrum sensing. The main goal of such protocols is to establish and maintain route across region of different available spectrum. In the rest of the sub section we summarize these protocols and highlight their advantages and their routing techniques. Protocol summery in the form of table is presented at the end of each protocol.

4.1.1 SPEctrum-Aware Routing (SPEAR)

A routing protocol that supports high-throughput packet transmission in the presence of spectrum heterogeneity is being investigated in [10]. It achieves persistent end-to-end performance by integrating flow-based approaches with link-based approaches. It assigns different channels to links on the same flow for minimizing interference and integrates spectrum discovery with route discovery for optimal usage of available channels. For this, each node maintains a list of unoccupied locally available channels. These channels are neither occupied by primary user nor reserved by nearby neighbors. In SPEAR route discovery is done by broadcasting a Route Request (RREQ) message on common control channel and being identified by sender and receiver IP addresses. As an intermediate node receives this message, it checks if it has a common channel with the previous node then it appends its own id and available channel set with the received message and then broadcasts it. The destination node selects the best path on the basis of maximum throughput, minimizes end-to-end latency (minimum hop count) and link quality. During the transmission node periodically broadcasts channel reservation message with each message containing timeout and time-to-live field. At the end of communication nodes along the path are notified to stop sending reservation messages.

4.1.1.1 Advantages and Overhead

SPEAR distributed on demand routing protocol that utilizes the spectrum heterogeneity. It has low computational and communication complexity. It almost achieves end-to-end throughput near optimal and adding additional nodes does not degrade its performance. Timeout field introduced in route set up simplifies management and provides robustness against link failure. Similarly time-to-live field minimizes broadcast overhead and contention. This protocol can be used with the narrow band control channel. On demand route discovery measured in term of route set up and route tear down delay and periodic channel reservation messages involves additional overhead.

4.1.2 Spectrum Aware MEsh Routing (SAMER)

A routing protocol for mesh CRNs proposed in [11] handles the diversity in channel availability and balance between long-term route stability and short-term route. SAMER uses the available white spaces by transmitting the data over the route with higher spectrum availability. Thus spectrum availability is used for computing routing metric for long-term routes. It achieves the balance between long- and short-term routes by constructing a runtime forwarding route mesh. This mesh is periodically updated and provides a set of candidate routes to the destination. Thus packets are routed towards the destination across this mesh. The routing decisions are taken with the collaboration of PHY and MAC layer. SAMER builds dynamic candidate, candidate forwarding mesh and opportunistically forwarding.

4.1.2.1 Dynamic Candidate Mesh:

In constructing dynamic candidate mesh a cost to destination is computed by every node in the network. The cost actually shows the spectrum availability of the highest spectrum route. This route contains fewer hops than a specified threshold. Thus each node constructs a set of forwarding nodes to destination.

Opportunistic Forwarding: SAMER uses those links for forwarding that contains the highest spectrum availability. It used PSA metric which is defined in term of throughput between a pair of nodes for computing spectrum availability.

4.1.2.2 Building a candidate forwarding mesh:

Forwarding mesh only contains long-term paths. These paths are the shortest in term of hope count. Thus for forwarding packet to destination D a node i computes a cost for all of its neighboring nodes and it adds only those nodes in its forwarding set that contain minimum cost. Therefore a cost to destination is prior computed by each network node.

4.1.2.3 Advantages:

SAMER achieves high end-to-end throughput by using long-term stability and short-term opportunistic utilization of spectrum. Changes in spectrum availability are adopted dynamically by its forwarding mesh.

Table 1: Summary of SPEAR& SAMER Routing Protocols.

	SPEAR		SAMER	
Objective and features		**Technique**		**Technique**
End-to-end throughput	Yes	Integration of flow-based and link-based approaches	Yes	Path with high spectrum availability, long-term stability and short-term opportunistic utilization of spectrum
Route discovery		Control channel, Broadcasting RREQ message		Link state packets
Routing decisions		With the Collaboration of PHY and		With the Collaboration of PHY and MAC layers

		MAC layers		
Route nature		On demand		Periodical
Mobility handling	Yes	Timeout field in periodic channel reservations messages	No	
Communicati on overhead	Yes	Route setup, route tear down	Yes	Each node computes its cost to destination,
Computation al Complexity	Low		Low	
Route discovery packet size	Variabl e	Each node appends its identifier and available channel set	Fixed	Hop-to-hop calculation
Best path selection	Yes	Maximum throughput, minimum hop count and link quality	Yes	Minimum hop count and spectrum availability

4.1.3 Spectrum-aware On-Demand routing protocol (SORP)

SORP is an on demand routing protocol that is neither based on centralized spectrum allocation nor multi-channel. The nature of this protocol is due to lack of shared information. The routing technique proposed by Cheng et.al in [12] is to select best suitable RF bands for each node along the route. The RF band selection is based on minimum cumulative delay. The switching and back-off delay caused by both the path itself and the intersecting flow are the judging parameters for calculating cumulative delay of the path. They proposed a spectrum aware on demand framework for routing and multi-flow multi-frequency scheduling for RF band selection. They slightly modified Ad hoc on demand distance vector routing (AODV) [20] to incorporate the inconsistency of spectrum opportunity. They made some assumptions for their routing technique, as follow:

To form a common control channel each node contains a traditional wireless interface in addition to the CR transceiver. Each node is able to provide spectrum sensing information to routing protocol through cross layer design.

For route discovery SORP inherits the basic procedures of AODV with modified Route Request (RREQ).In SORP Spectrum Opportunity (SOP) information is piggybacked by RREQ messages. SOP information is piggybacked only when the node finds intersection between the RREQ and its own. Thus destination node receives the SOP distribution of all the nodes along the path and it assigns RF band to its CR transceiver accordingly. This RF band information is sent back to the source node as well intermediate nodes through Route Reply (RREP) message. All the nodes along the path assign the RF band according to the received RREP.

4.1.3.1 Advantages and Overhead:

This routing technique overcomes the inconsistency of SOPs. It selects best path on the basis of the total delay along the path. It selects the RF band with joint interaction of routing and scheduling. Therefore path cumulative based RF band selection introduces both switching and back off delay.

4.1.4 Multi-hop Single-transceiver Cognitive Radio Networks Routing Protocol (MSCRP):

MSCRP proposed in [13] doesn't base on control channel. Therefore, routing protocol messages are being exchanged without common control channel. MSCRP is an on demand protocol based on ad hoc on demand distance vector (AODV).

Ma et.al modifies AODV to handle the available channel set problem that each node in the network doesn't know the available channel set of other nodes in the network. In [13] they first time introduced the new problem called "deafness", that is due to channel switching of the nodes. To avoid the deafness problem, they proposed that two consecutive nodes in a flow cannot be in the switching state simultaneously. Communicating with a switching node is complicated, therefore MSCRP switching node uses LEAVE/JOIN messages to inform its neighbors about its working channel. MSCRP assumes that CR transceiver can tune in a wide range of RF spectrum but it only operates on limited and smaller range of RF and CR transceiver can only operate on single channel at any time. MSCRP is a cross layer protocol so it identifies six system functions that implement the core functionality of spectrum aware routing. These functions are as follow:

The physical layer includes three of them that are spectrum sensing, detecting active primary user and estimating the quality of available channels.

The network layer includes two of them that are routing and scheduling in the multi-flow and multi-channel environment.

Link layer has last one that is IEEE 802.11DCF is used as the MAC protocol.

4.1.4.1 Route Discovery Mechanism in MSCRP:

RREQ message is broadcast on all the channels for route discovery. The channels availability information piggybacked by RREQ a message is forwarded in broadcast process. All intermediate nodes append their state and available channel set to RREQ message. As nodes may stay on different channels, therefore broadcast manner is totally different from original broadcast used in AODV. The reverse path to the source node is established as RREQ is forwarded. Destination node receives the channel information and number of nodes on each channel at the end and assigns channel for this flow. It encapsulates the assigned channel information in RREP message.

4.1.4.2 Advantages and Overhead:

It first time introduces the deafness problem and introduces a novel approach to deal it. This approach is well fit in multi-hop and single transceiver CRNs. Dealing with single transceiver is cost-effective as compared to multi-transceiver. The deafness introduces some extra delay to RREQ messages due to its channel switching. MSCRP also introduces extra overhead of broadcasting RREQ message on all available channels rather on single channel and this overhead will becomes insufferable in case many available channels on each network node.

Table 2: Summary of SORP & MSCRP Routing Protocols.

	SORP		MSCRP	
Objective and features		Technique/ explanation		Technique/ explanation
End-to end throughput	Yes	spectrum aware on demand routing and multi-flow multi-frequency scheduling	Yes	Spectrum aware routing and leave/join messages
Route discovery		Broadcast RREQ messages		RREQ message on all available channels rather on single channel
Routing decisions		With the collaboration of MAC and network layers		With the collaboration of MAC, physical and network layers
Route nature		On demand		On demand
Link failure handling	No		Yes	Leave/join messages
Communication overhead		Path cumulative based RF band selection introduces both switching and back off delay		Deafness introduces extra delay to RREQ due to its channel switching and broadcasting RREQ messages on all available channels introduce extra overhead
Route discovery packet size	Variable	Each node appends SOP list	Variable	All intermediate nodes append their state information and available channel set
Best path selection	Yes	Path delay and node delay (switching and back off delay)	Yes	Number of flows on each channel
Control channel	Yes	Exchanging the routing protocol messages	No	Data channel for routing protocol messages

4.2 Reactive Source-Based Routing

In reactive source based routing technique source specifies how the data travels across the network. Path to destination node is computed by the source node. In the rest of the sub section we summarize a reactive source-based routing protocol and highlight its routing technique and its advantages.

4.2.1 Routing in Opportunistic Cognitive Radio Networks

Reactive source-based routing protocol for CRNs is proposed by Khalife et.al [14] and it uses a novel routing metric that is based on a probabilistic definition of the available capacity over a channel. This routing metric determines the most probable path (MPP) to satisfy a given bandwidth demand although it doesn't guarantee to satisfy the demand. So in this case an augmentation phase is used in which bottleneck links are augmented with additional channels so the resulting path meets the bandwidth demand with a given probability.

The available capacity is measured as the probability distribution of the PR to CR user interference at any node over a channel.

When an application requests a route of capacity demand the source will initiate it and control channel is used for node coordination. Based on the demand all links probabilities are calculated. Once all link weights are calculated, the source runs Dijkstra-like algorithm to find a route to the destination. The obtained path is called MPP as it has the highest probability of satisfying the demand and stability to destination. The Dijkstra-like algorithm stops computing when it reaches to the one of the following two states.

1. On each link of MPP, the total capacity will be greater than the demand.
2. After augmentation if the total estimated capacity on all the channels of two nodes will not fulfill the demand. In this case no path is suitable to the destination thus it is declared unreachable.

4.2.1.1 Advantages:

This routing protocol deals with the simultaneous transmissions over multiple channels as well Primary Radio-to-Cognitive Radio interference.

Table3: Summary of Reactive Source based Routing Protocol.

Objective and features		Technique /explanation
End-to-end throughput	Yes	Selecting MPP path that fulfils the application capacity demand
Route discovery		OSPF, Dijkstra-like algorithm to compute the route
Routing decisions		Does not base on cross layer
Route nature		On demand
Data structure		Graph
Link weight assignment	Yes	Based on available capacity of link
link failure handling	No	
Route discovery packet size	Fixed	Hop-to-hop communication
Best path selection	Yes	A routing metric based on the probabilistic definition of available capacity over channel
Control channel	Yes	

4.3 Local Coordination-Based Routing

The local coordination is a sort of enhancement scheme that is applied on intersecting nodes on a path. The local coordination is started when nodes evaluate the workload of both accommodating the flow and redirecting it. Nodes choose the flow accommodation or flow redirection based on the evaluation results and neighborhood interaction.

In the rest of the sub section we summarize a local coordination-based routing protocol and highlight its routing technique and its advantages.

4.3.1 Local Coordination Based Routing and Spectrum Assignment in Multi-hop Cognitive Radio Networks:

An on demand routing and spectrum assignment protocol to exchange the local spectrum information and interact with multi-frequency scheduling in each node is proposed by Yang et.al [15]. AODV is modified to form a mechanism on common control channel for exchanging spectrum opportunity (SOP) among the nodes to overcome the inconsistency of SOP. It also identifies traversing flows at every node and calculates RF band used by any node and this is used for multi-flow multi-frequency scheduling. Path delay and node delay show the switching and back off delays along the path and used to calculate the cumulative delay of the path. A local coordination scheme is used for load balancing on intersecting nodes for multi-frequency traffic. Each network node is equipped with traditional wireless interface in addition to CR transceiver to ensure the successful delivery of routing messages at each node despite of the inconsistency of the frequency bands as well every node provides the SOP information to its network layer. The local coordination is applied on every network node of multi-hop CRNs.

4.3.1.1 Advantages:

The proposed routing protocol provides good adaptability for spectrum diversity and end-to-end delay. This scheme outperforms the traditional bare routing.

Table 4: Summary of LCB Routing Protocol.

Objective and features		Technique /explanation
End-to-end delay	Less	Adaptive relay is cooperating with routing protocol
End-to-end performance		Redirecting flow to other neighboring nodes and accommodating the flow
Route discovery		Broadcast RREQ messages
Routing decisions		Joint decisions based on MAC and Network layer
Route nature		On demand
Link failure handling	Yes	Redirecting flow to another neighbor
Protocol overhead		
Route discovery packet size	Variable	Each intermediate node appends SOP list
Best path selection	Yes	Based on the cumulative delay of the path
Load balancing	Yes	A local coordination scheme is used
Network topology		Full mesh
Control channel	Yes	Exchanging spectrum opportunity among network nodes
Neighbor discovery	Yes	Channel scanning and beacon broadcast

4.4 Multi-path Routing

In multi-path routing multiple routes are discovered for any destination and then some best routes among discovered route are selected based on different parameters. Multi-path routing has many benefits such as fault tolerance, increased bandwidth and reduction of primary to secondary user interference. In the rest of the sub section we summarize a multi-path routing protocol and highlight its routing technique and its advantages.

4.4.1 Multipath Routing and Spectrum Access (MRSA)

Existing multi-path routing protocols for traditional wireless networks cannot be adapted in CRNs since they neither consider the diversity in spectrum availability nor coexistence of primary and secondary users. MRSA [16] is the first multi-path protocol for CRNs that minimizes the inter path contention and interference. It overcomes the interruption of primary users with minimum degradation by distributing the traffic of each flow over multiple paths.

For traffic distribution it uses round robin fashion that is not an effective technique. In MRSA "spectrum wise disjointness" concept is revised as if multiple paths do not have any interfering bands between them then these paths are spectrum wise disjointed. MRSA assumes that there will be total N channels for data traffic and signaling is delivered over these channels together with data traffic. It uses dynamic source routing (DSR) [21] mechanism for route discovery in which source node broadcasts an RREQ message with new RREQ_ID and attaches its band radio usage table (BRT). When an intermediate node receives RREQ before forwarding, it verifies if the RREQ_ID is new or if RREQ_ID is not new then it counts the hop count from source. If RREQ has fewer hop count than the previous RREQ it will append its BRT and then forwards it. Thus destination will receive the same RREQ from multiple paths. Thus it first assigns band and radio to each link then evaluates all the candidate paths by their available bandwidth. RERR message of DSR is extended to overcome the sudden arrival of primary user and it's the part of route recovery process.

4.4.1.1 Advantages:

The protocol constructs multiple paths to maximize spectrum wise disjointedness and to minimize contention and interference. It achieves higher throughput than other routing approaches and effectively utilizes the network resources. It also provides better resilience from the dynamic interruption of primary users.

Table 5: Summary of MRSA Routing Protocol.

Objective and features		Technique /explanation
End-to end throughput	Yes	Using multi-radios and multipath
Route discovery		Control channel, Broadcasting RREQ message
Routing decisions		Does not base on cross layer
Route nature		Periodical
Path failure handling	Yes	Used RERR messages
Communication overhead	Yes	Multiple flows on single radio therefore band switching for each individual flow
Route discovery packet size	Variable	Each node appends its ID and BRT
Best path selection	Yes	Minimum hop count

Data striping	Yes	Round robin fashion
Network topology		Mesh

4.5 Tree Based Routing

In tree based routing protocol a tree structured network is enabled by configuring a root. Tree based routing is centralized routing scheme which is controlled by a single network entity called base station. Thus network topology can be quickly constructed among CR station by configuring cognitive base station as root. In the rest of the sub section we summarize a tree based routing protocol and highlight its routing technique and its advantages.

4.5.1 Cognitive Tree-based Routing (CTBR)

Cognitive tree based routing (CTBR) [19] is an extension of tree based routing protocol (TBR) proposed for wireless mesh networks [17-18]. It uses global and local decision schemes for route calculation. Global decision scheme selects route with the best global end-to-end metric whereas local decision scheme selects the best interface with the least load. Multiple paths with the same global end-to-end metric can exist for the same destination. In this case the end-to-end path is selected based on the local decision scheme, which uses load measuring.

CTBR uses the routing procedure of TBR in which root periodically sends Root Announcement (RANN) message for tree formulation. Any node receives the RANN, caches the node whom it receives the RANN as its potential parent and then rebroadcast RANN with updated cumulative metric. The node will select a parent node from all potential parents based on the best metric (i.e. hope count) for the path to root. For registering with root every node that contains known path to root sends route reply (RREP). Any intermediate node that receives RREP forwards the message to its parent node as well updates its routing table by selecting source node of RREP as its destination. Thus at the end root constructs a tree as it has learnt all network nodes. To make TBR adaptable for CRNs a link quality metric has been introduced.

4.5.1.1 Advantages:

In this routing protocol they proposed a new cognitive aware link metric for computing global end-to-end metric.

Path is selected on the basics of global and local decision schemes. Average end-to-end delay of CTBR achieves about 5 times smaller than any hop count based scheme.

Table 6: Summary of CTBR Routing Protocol.

Objective and features		Technique /explanation
End-to-end delay		5 times better than hop count scheme
Route discovery		Broadcast Root Announcement (RANN) message
Routing decisions		Does not based on cross layer
Route nature		Periodical
link failure handling	No	
Protocol overhead	Yes	Additional control bytes are transmitted
Route discovery packet size	Fixed	Every node updates a single field known as "cumulative metric"

Best path selection	Yes	Path is selected on the basics of global and local decision schemes
Centralize	Yes	Single point of failure
Network topology		Tree

5. CONCLUSION

Routing in multi-hop CRNs is challenging task due to the diversity in channel availability and data rates, therefore currently researchers from all around the world are focusing to introduce some novel routing techniques for CRNs. In this paper, firstly we discussed the architecture and main routing differences and challenges for CRNs then we presented a comprehensive survey and analytical analysis of available routing techniques for CRNs. The routing techniques are classified into dynamic spectrum aware-based, multipath-based, local coordination-based, reactive source-based and tree-based depending on their protocol operation. We also highlighted the routing operation, as well as the advantages and disadvantages of each routing technique. Although many routing techniques look promising but mostly presented techniques uses the same routing metrics as conventional wireless networks. Therefore, there is need to design new metrics those exploits all the dynamic characteristics of CRNs and based on such metrics novel routing proposals should be presented.

REFERENCES

[1]   B. Fette, Cognitive Radio Technology, Elsevier Inc. (2006).

[2]   Yuan Yuan, Paramir Bahl, Ranveer Chandra, Thomas Moscibroda, Yunnan Wu. Allocating Dynamic Time-Spectrum Blocks In Cognitive   Radio Networks. In ACM MobiHoc 2007.

[3]   Federal Communications Commission, Spectrum Policy Task Force Report, ET Docket No. 03-222, Notice of Proposed Rule Making and Order2003.

[4]   I. F. Akyildiz, W.-Y. Lee, M. C. Vuran, and S. Mohanty, NeXt generation/ dynamic spectrum access/cognitive radio wireless networks: A survey, Elsevier Computer Networks, 50(3), 2127–2159 (Sept, 2006).

[5]   R. W. Thomas, L. A. DaSilva, and A. B. MacKenzie, "Cognitive networks," in Proc. of IEEE DySPAN2005, pp. 352–360, November 2005.

[6]   Yang Xiao, Fei Hu "Cognitive radio networks". 2009 Taylor & Francis Group, LLC. ISBN: 978-1-4200-6420-9

[7]   De Cout, D. S. J, et.al. A high-throughput path metric for multihop wireless routing. In Proc. of MobiCom (Sept. 2003).

[8]   Ma, H. and Zheng, L. and Ma, X. and Luo," Spectrum-aware routing for multi-hop cognitive radio networks with a single transceiver", Proceedings of the Cognitive Radio Oriented Wireless Networks and Communications (CrownCom) 2008

[9]   Kwang-Cheng, Chen, Ramjee Prasad, "Cognitive radio networks". 2009 John Wiley & Sons Ltd. ISBN: 978-0-470-69689-7

[10]   Sampath, A. and Yang, L. and Cao, L. and Zheng, H. and Zhao," High Throughput Spectrum-aware Routing for Cognitive Radio Networks.

[11]   Pefkianakis, I. and Wong, S.H.Y. and Lu," SAMER: Spectrum Aware Mesh Routing in Cognitive Radio Networks", 3rd IEEE Symposium on New Frontiers in Dynamic Spectrum Access Networks, 2008. DySPAN 2008.

[12]   Cheng, G. and Liu, W. and Li, Y. and Cheng," Spectrum aware on-demand routing in cognitive radio networks", 2nd IEEE International Symposium on New Frontiers in Dynamic Spectrum Access Networks, 2007. DySPAN 2007.pp.571—574.

[13]   Ma, H. and Zheng, L. and Ma, X. and Luo," Spectrum-aware routing for multi-hop cognitive radio networks with a single transceiver", Proceedings of the Cognitive Radio Oriented Wireless Networks and Communications (CrownCom) 2008.

[14]   Khalife, H. and Ahuja, S. and Malouch, N. and Krunz," Routing in Opportunistic Cognitive Radio Networks".

[15]   Yang, Z. and Cheng, G. and Liu, W. and Yuan, W. and Cheng."Local coordination based routing and spectrum assignment in multi-hop cognitive radio networks", Mobile Networks and Applications 2008.

[16]   Wang, X. and Kwon, T.T. and Choi," A multipath routing and spectrum access (MRSA) framework for cognitive radio systems in multi-radio mesh networks" Proceedings of the 2009 ACM workshop on Cognitive radio networks 2009.

[17]   IEEE 802.11s http://www.802wirelessworld.com/.

[18]   A. Raniwala and T. C. Chiueh, "Architecture and algorithms for an IEEE802.11-based multi-channel wireless mesh network," in Proc. IEEE INFOCOM Conf., pp.2223-2234, 2005

[19]   B. Zhang, Y. Takizawa, A. Hasagawa, A. Yamauchi, and S. Obana, *Tree-based routing protocol for cognitive wireless access networks,"* in Proc. of IEEE Wireless Communications and Networking Conference2007.

[20]   N1: C. E. Perkins and E. M. Royer, "Ad hoc on-demand distance vector routing," in Proc.of IEEE Workshop on Mobile Computing Systems and Applications, 1999.

[21]   N2: D. B. Johnson, D.A. Maltz and Y. C. Hu, "The Dynamic Source Routing for mobile ad hoc networks,", draft-ietf-manet-dsr-09.txt, 2003.

13

IMPROVING AODV PERFORMANCE USING DYNAMIC DENSITY DRIVEN ROUTE REQUEST FORWARDING

Venetis Kanakaris, David Ndzi, Kyriakos Ovaliadis

Department of Electronic and Computer Engineering,
University of Portsmouth, Anglesea Road, Portsmouth, PO1 3DJ, United Kingdom
www.port.ac.uk
Email: {venetis.kanakaris,david.ndzi,kyriakos.ovaliadis}@port.ac.uk

ABSTRACT

Ad-hoc routing protocols use a number of algorithms for route discovery. Some use flooding in which a route request packet (RREQ) is broadcasted from a source node to other nodes in the network. This often leads to unnecessary retransmissions, causing congestion and packet collisions in the network, a phenomenon called a broadcast storm. This paper presents a RREQ message forwarding scheme for AODV that reduces routing overheads. This has been called AODV_EXT. Its performance is compared to that of AODV, DSDV, DSR and OLSR protocols. Simulation results show that AODV_EXT achieves 3% energy efficiency, 19.5% improvement in data throughput and 69.5% reduction in the number of dropped packets for a network of 50 nodes. Greater efficiency is achieved in high density network and marginal improvement in networks with a small number of nodes.

KEYWORDS

Mobile Ad-hoc Network, Routing Message Overhead, Route Discovery, Broadcast, flooding, Power Aware Routing, Routing Protocols, AODV_EXT, AODV, DSDV, DSR, OLSR.

1. INTRODUCTION

A combination of centralized and ad-hoc networks is envisaged to provide solutions for the provision of ubiquitous communication for a wide range of applications. Whilst centralized communication is well established, ad-hoc networking is seen as the way forward for self organizing and managing networks which eliminate meticulous and expensive planning, high cost, rigidity and vulnerability inherent in fixed centrally managed networks such as wired and wireless networks e.g. Mobile Telecommunications System (UMTS). Ad-hoc networks also hold great promise and applications in an extensive number of areas ranging from disaster management to environmental monitoring. Progress in ad-hoc networks is also facilitating the application of sensors for process automation in a variety of industries and is enabling progress in sensor fusion. Unpredictable events, e.g. earthquake often serve to illustrate the vulnerability of centrally managed networks and the importance of research and development in ad-hoc networks such as Mobile Ad-hoc Networks (MANET), for which centralized connectivity is not required. MANET is a wireless network that has mobile nodes with no fixed infrastructure. The main limitation of ad-hoc network systems is the availability of power and continuous reduction in the size of devices mean that power limitation cannot simply be ameliorated with large battery packs [1] . In addition to running the onboard electronics, power consumption is governed by the number of processes and overheads required to maintain connectivity.

A wide range of techniques have been proposed to address connectivity and power limitation issues in ad-hoc networks. These include hardware development, protocols, routing algorithms and battery technology or energy management systems [2] . Some researchers have proposed the development of hardware optimized for specific applications based on data rates [3] . More detailed studies of energy consumption by hardware have been carried out to evaluate energy consumption when transmitting, receiving, in sleep and idle modes. It has been proposed in [4] [5] that energy management should be tailored to each application where the voltage, and hence processing speed and energy, can be reduced for non-time sensitive applications. This proposed technique benefits from the fact that the speed of microprocessors and energy consumption depend on the voltage that is applied to it. The common goal of energy management techniques proposed for ad-hoc networks is to preserve energy and maximize life span of the network. Other proposed methods are aimed at preventing network partitioning by managing energy consumptions of critical link nodes. In this paper we propose a modification of one of the most widely used protocols to improve the energy and data transmission efficiency of the network.

In general, ad-hoc wireless networks broadcast packets to the whole network as a means of transmitting information from one node to the other in the network [6] . Broadcasting in MANETs is not only a fundamental action for unicast routing protocols in mobile scenarios, but it is also an inextricable part of a number of multicast routing protocols. There is a diversity of geocast, unicast, and multicast protocols that use the broadcasting procedure in order to provide the significant function of control and route establishment. Broadcasting a packet to the entire network has extensive applications in mobile ad-hoc networks. Therefore, improving the broadcasting process will result in savings in several MANET applications.

Flooding is the simplest technique used by source nodes to broadcast packets to neighbouring nodes [7] . Each neighbour node receiving the packet for the first time rebroadcasts it ensuring outward propagation from the source until every node in the network has received and transmitted the broadcast packet exactly once.

Significant research activities have focused on reducing flooding in the network [7] [8] . Any procedure that leads to a reduction in congestion saves energy and prolongs the life span of the network. In general, multi-hop transmissions are less energy efficient because of the startup energy consumption of the transceivers [9] . Therefore flooding which results in the reception and retransmission by multiple nodes in a network where path loss is not the dominant energy consumption element is energy inefficient. In high density networks allowing nodes to be turned off or enabling sleep mode and maximizing transmission range can increase energy efficiency. In [10] the concept of a minimum range routing where nodes within a specific range of a transmitting node are not allowed to retransmit a packet has been proposed. However, the proposed technique relies on nodes keeping an updated table of information about neighbouring nodes which can be time and energy inefficient in a high mobility environment. Another approach has been proposed in [11] where power consumption is distributed amongst the nodes by controlling the transmission and the reception powers. Using this technique, the amount of power consumed for sending one packet to any destination node is the same and determined for each node that is taking part in the routing process.

Broadcast protocols can be broadly divided into two main categories; deterministic and probabilistic. The probabilistic approach usually provide a simple solution in which every node that receives a broadcast packet has a fixed probability of forwarding the message [10] . But this approach does not guarantee full network coverage. On the other hand the deterministic approach can provide full coverage and can be further grouped into two categories, location information and neighbour set based.

In MANETs the routing task is delivered through network nodes which act as both routers and end points in the network. In order for a route to a specific destination node to be discovered, existing on-demand routing protocols use a simple flooding mechanism whereby a Route Request packet (RREQ) originating from a source node is broadcasted without exception to all nodes in the network [12] . This can lead to significant redundant retransmissions, causing high channel usage and packet collisions in the network.

In this paper there will be presented how the broadcasted routing messages react on the network performance. The protocols Ad-hoc On-Demand Distance Vector Routing (AODV), Dynamic Source Routing (DSR), Destination Sequenced Distance Vector (DSDV) routing and Optimized Link State Routing (OLSR) have been studied thoroughly and their performances in simulated networks has been analysed. These protocols have been widely used and cited in literature [13] .

Types of MANET Routing Protocols

One of the main tasks of routing protocols is to maintain routes inside MANET, since they do not use any access points to connect to other nodes in the network [14] . Routing protocols can be classified into three categories depending on their properties as follows:

➢ Centralized or distributed
➢ Static or adaptive
➢ Reactive or proactive

In centralized networks all route choices are made by a central node whilst in distributed routing networks the computation of routes is shared amongst the network nodes. In static routing, the route used by source and destination pairs is fixed regardless of traffic condition. It can only change according to the node needs or link failure. This kind of algorithm cannot get high throughput in cases of traffic conditions variety. In adaptive routing, the routes used between source and destination pairs may change in response to traffic condition e.g. congestion. There is another classification that the ad-hoc networks can be classified regarding the routing algorithms: proactive or reactive.

In proactive routing protocols, nodes maintain one or more routing tables about nodes in the network. These routing table information are updated either periodically or in response to a change in the network topology. The advantage of these protocols is that a source node does not need route-discovery procedures to find a route to a destination node. The disadvantage of proactive routing protocols is that because of it keeps an up-to-date routing table, creates essential messaging overhead, which consumes energy and bandwidth, and reduces throughput, especially when there is a large network with high node mobility. There are various types of table driven protocols which include: DSDV, OLSR, Wireless Routing Protocol (WRP) [15] , Fish eye State Routing (FSR), Cluster Gateway Switch Routing (CGSR) protocol, and Topology Dissemination Based on Reverse Path Forwarding (TBRPF) protocol [16] .

For reactive (on-demand) protocols there is an initialisation of a route discovery mechanism by the source node to find the route to the destination node when the source has data packets to send. When a route is found, route maintenance process is initiated to maintain this route until it is no longer required or the destination is not reachable. Reducing the message overhead is the advantage of these protocols. One of the drawbacks of these protocols is the delay in discovering a new route. Examples of reactive routing protocols include DSR, AODV and Temporally Ordered Routing Algorithm (TORA) [15] .

This paper is organized as follows; Section 2 describes the protocols that will be evaluated in this paper. The routing procedure of AODV is described in more details and this is followed by the description of the proposed modification to AODV. Section 3 describes the simulated scenario, the settings, network configurations and the parameters that have been used to assess the performance of the protocols. The results of the simulation and discussions are provided in Section 4 which is followed by the conclusions.

2. ROUTING PROTOCOLS

For DSDV [17]   protocol messages are exchanged between mobile nodes within range. Routing updates may be triggered or routine. Updates are initiated when routing information from one of the neighbours forces a change in the routing table. If there is a packet which the route to its destination is unknown, it is cached while routing queries are sent out. The data packets are stored temporarily until the destination node receive route-replies. The buffer has a size and time limit for caching packets beyond which packets are dropped. All packets for which the route to their destinations is known are routed directly. In the event that a target is not found, the packets are forwarded to the default target which is the routing agent. The routing agent designates the next hop for the packet.

On DSR protocol the agent node controls every data packet regarding the information of source-route [18] . The packets are then forwarded as per the routing information. If there is no routing information in the packet, it gives the source route if route is known. When the destination is not known it caches the packet and sends out route queries. The routing query is initially sent to all nearby nodes and is always triggered by a data packet which has no route information regarding its destination. Route-replies are sent back if routing information to the destination is found.

AODV protocol is a mixture of both DSR and DSDV protocols [19] . It has the mechanism of route-discovery and route-maintenance of DSR. Moreover it keeps from DSDV the hop-by-hop routing sequence numbers and beacons. When a node needs to know a route to a specific destination, it creates a RREQ. The RREQ is forwarded by intermediate nodes which also create a reverse route from the destination. When the request reaches a node with a route to the destination node it also creates a Route Reply (RREP) which contains the number of hops that are required to reach the destination. All nodes that participate in forwarding this reply to the source node create a forward route to destination. This route is made not from complete route as in source routing but from every node which is a hop-by-hop state from source to destination.

OLSR [20]   is a routing protocol where the nodes know all the available routes. As an optimized version of the pure link state protocol, the OLSR protocol floods the topological information to all active nodes in the network when the topology changes. A way to reduce the possible overhead in the network protocol is to use Multi-point Relays (MPR). The idea of MPR [21] is to diminish the number of duplicate retransmissions when a broadcast packet is forwarded. In this technique the number of retransmissions is restricted to a small set of neighbouring nodes, instead of using all the neighbours. This set is kept as small as possible by choosing the nodes which cover (in terms of one-hop radio range) the same network region as the complete set of neighbours. The OLSR routing protocol uses two types of control messages; Hello and Topology Control (TC). Hello messages are used in order to find out information regarding the link status and the host's neighbours. On the other hand TC messages are used for sending information about neighbours which includes the MPR selector list. The OLSR protocol has a disadvantage in that every host periodically sends the updated topology information to the entire network thus increasing the bandwidth usage. But this issue is solved by using the MPR, which forwards only the messages regarding the topology of the network.

AODV Process

Normal RREQ and RREP processing mechanism of AODV is as follows:
- ➤ The source node S tries to send a packet to destination D.
- ➤ If S does not know the next hop for D, then it broadcasts a route request message.
- ➤ The RREQ message propagates in all directions to reach the destination D.
- ➤ The intermediate nodes that receive the RREQ message forward the packet to all its one hop adjacent nodes.
- ➤ If the destination, D, receives a RREQ message through a node N, then it sends a RREP to S by forwarding it to N since N may contain at least one routing table entry for S.
- ➤ On receiving the RREQ message through different nodes, the destination D will send the RREP message through different nodes and they may reach the source node through different possible paths.
- ➤ At the end, the source node S will have different possible resolved paths to select from based on defined criteria.

Proposed modification of AODV RREQ mechanism

Standard AODV routing process broadcasts route request to all nodes. In the proposed scheme, a table of nodes in a given neighbourhood (one-hop nodes) is maintained. When a message is transmitted, only a subset of nodes in each neighbourhood is allowed to transmit. The number of selected nodes can be varied dynamically depending on the application and required quality of service. In this proposed scheme, the parameters that are used are defined in Table 1. Each node in the network will forward a route request message if and only if a condition based on its neighbourhood density at that instance is satisfied. The proposed scheme minimizes network congestion due to redundant transmission.

Table 1: Definition of AODV_EXT Parameters

n	The total nodes in the network
F_i	Any node F_i, i = 1,2,...n that receives the RREQ message
P_i	Packet forwarding probability derived from neighbour node count.
β_i	The number of nodes neighbouring node F_i.
d	Minimum number of neighbouring nodes - if the number of neighbours at a forwarding node, F_i, is less than or equal to d, then that node will forward the RREQ message to avoid path failure or network partitioning.
C_f	It is a control factor which can be used to adjust the probability P_i according to the application or average expected node density of the network, ($0 < C_f \leq 1$).
R	Random number (between 0 and 100). This is used to generate varying conditions in the network.

If the RREQ is received from an intermediate node then there will be at least one possible path which includes that node in its path list. Therefore, if only selected nodes are allowed to forward the RREQ packet, then only these nodes will be included in the path list. In this proposed scheme, the neighbourhood density of an intermediate node is considered as a criterion in RREQ forwarding decision at intermediate node. It means that if the number of nodes in the neighbourhood is high, then the probability of any node transmitting will decrease and hence reduces the transmission overhead. Random selection of nodes from the neighbourhood set increases the chance of full network coverage. Greater savings could be achieve by using a range dependent technique to select nodes for transmission but this can only be achieved at the cost of greater complexity.

PROPOSED ALGORITHM

Any node $F_i, i = 1, 2, ..., n$ receiving the RREQ message will process the packet as follows :

For RREQ message originating from S destined for node D that is received by node F_i process it if $F_i \neq S$ and $F_i \neq D$ (i.e. F_i is an intermediate node) as follows:

Node F_i resolves its neighbourhood density β_i

If $\beta_i \leq d$ then

        Forward the RREQ message

Else

        Calculate message forwarding probability P_i at node F_i

$$p_i = \frac{100}{\beta_i} * (d * C_f) \qquad \text{for } 0 < C_f \leq 1$$

        If $R < p_i$ then

                Forward the RREQ message

        Else

                Ignore and Drop the RREQ message

        End

End

---

3. SIMULATION AND METRICS

The simulation set up assumes the use of 802.11 standards based on the two path propagation model described by equation (1).

$$P_r = P_t G_t G_r \left(\frac{h_t h_r \lambda}{4 \pi R^2} \right)^2$$

(1)

Where P_r and P_t are the received and transmitted power; G_r and G_t are the gains of the receiving and transmitting antennas; h_r and h_t are the heights of the transmitting and receiving antennas; λ is the wavelength of the signal and R is the distance between the transmitting and receiving nodes. This model assumes free-space. However, in reality the propagation conditions are usually more complex and often exhibit time and spatial variations resulting in shorter network life span than predicted by simulation. In addition, the simulation model takes an abstract view of the hardware power consumption by quantifying energy consumption based on the functions rather than hardware or changes in the propagation conditions. These assumptions are maintained in the simulation carried out in this paper because this can be accounted for by introducing a broad term for large scale variation. The energy model used can be described as follows [26] :

$$DecrTxEnergy = tx\_time * P\_tx \tag{2}$$

$$DecrRxEnergy = rx\_time * P\_rx \tag{3}$$

where DecrTxEnergy and DecrRxEnergy are the power consumed when transmitting and receiving, P_tx and P_rx are the power used when transmitting and receiving per unit time, respectively. The parameters tx_time and rx_time are the duration of transmission and reception, respectively. The lifetime of the network must take into account power consumed in three states: transmit, receive and idle. However, in this study power consumed in the idle state is not taken into account because the focus is on the effects of the protocols.

In the simulation configuration $d = 5$ was used. This defines a minimum condition which provides greater certainty of successfully packets routing but minimizes redundant transmissions. This is equivalent to minimum range routing but derived in terms of node density rather than distance.

Network Simulator 2 (NS2) has been used to evaluate the protocols [22] -[25] . The simulations were carried out to assess the performance of the routing protocols with network sizes of 10, 20, 30, 40 and 50 nodes with mobile node speeds between 1 m/s and 40m/s. For simplicity, in all cases the nodes send Constant Bit-rate (CBR) over User Datagram Protocol (UDP). The metrics that have been used to evaluate the performance of the network and protocols are the following [26] :

- *Number of packets dropped*: This is the number of data packets that are not successfully sent to its destination.
- *Consumed power*: The average consumed battery power.
- *Throughput*: This measure how fast the network can continuously send/receive data to the sink. Throughput is the number of packet received from the sink per millisecond.
- *MAC load*: This is the ratio of the number of MAC layer messages broadcasted from each node of the whole network to the number of data packets successfully delivered to all destination nodes. In other words, the MAC load is the average number of MAC messages generated for each data packet successfully delivered to the destination.
- *Control message overhead*: This control message overhead is the total routing control messages transmitted and received in the network.

The simulation configuration and specification are specified in Table 2.

4. RESULTS AND DISCUSSIONS

Figure 1 shows that AODV_EXT consumes less power than the other four protocols. Most importantly the power consumption of AODV_EXT based network improves in comparison to that of standard AODV as the number of nodes in the network increases beyond 30. The two protocols perform equally in small size networks. DSR protocol based networks consumes more energy compared to AODV and AODV_EXT but shows better performance when the number of nodes in the network is small (up to 30 nodes). On the contrary, the power consumption in networks using DSDV and OLSR rises steadily starting from fairly high levels. With increasing number of nodes, the energy depletion of OLSR based networks increases faster than those for the other protocols. OLSR protocol uses a mechanism that constantly updates information about nodes in the neighbourhood and therefore consumes more energy. As the number of nodes in the network increases, more updates are required and hence proactive protocols perform poorly, especially when the network is subject to changes e.g. in mobile environment.

Table 1: Simulation Parameters and Configuration

Routing Protocols	AODV, AODV_EXT, OLSR, DSDV, DSR
Topographical Area	800m x 800m
Number of Nodes	10, 20, 30, 40, 50
Mobility	1m/s to 40m/s
Channel type	Wireless Channel
Radio-propagation model	TwoRayGround
Network interface type	WirelessPhy
MAC type	802_11
Interface queue type	DropTail/PriQueue
Antenna model	OmniAntenna
Max packet in Queue	50
Transport /Traffic Type	CBR over UDP
TxPower of the nodes	0.1819 watts
RxPower of the nodes	0.0501 watts
IdlePower of the nodes	0.0350 watts
Initial energy of the nodes	1000.0 Joules

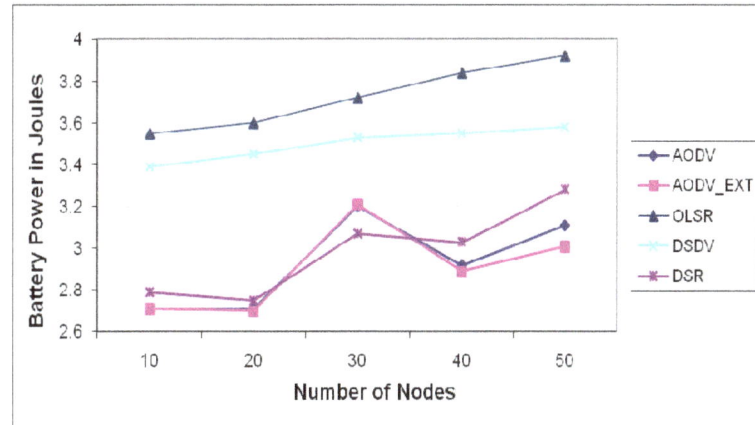

Figure 1: The Average Consumed Power

Figure 2 shows the performance of the protocols based on data throughput. It shows that AODV and DSR achieve comparable performance. However, AODV_EXT shows superior performance in larger networks. With AODV_EXT and AODV protocol, every node does not need to keep information regarding the route between two nodes. This reduces the amount of signaling required for route discovery and maintenance. OLSR and DSDV both show poor performances compared to the other three protocols. This is because both are proactive protocols and required table updates generate relatively high messaging overhead that can cause collision in large networks, especially in mobile networks, and reduces data rate performance of the network. However, these protocols are better suited to low data rate transmission because their self updating scheme ensures connectivity rather than the availability of bandwidth for application data.

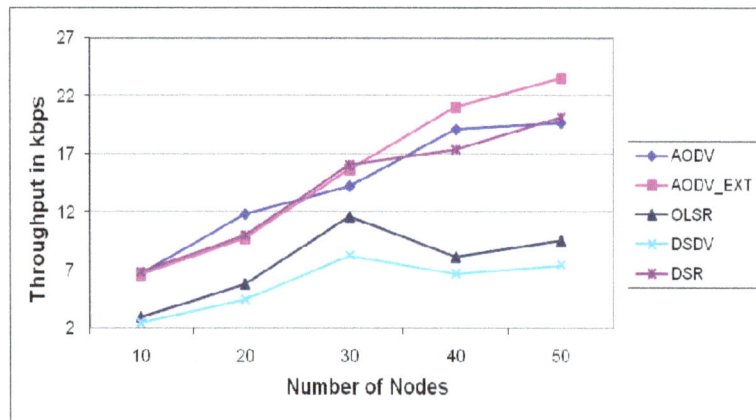

Figure 2: Comparison of data throughput for various network sizes (nodes)

Figure 3 shows MAC loading of the protocols. It shows that for large networks, a relatively high number of messages are generated by OLSR and DSR based networks. This increases sharply with the number of nodes in the network. DSDV and AODV exhibit only moderate increases. The figure shows that the use of density based scheme as applied in AODV_EXT significantly reduces the number of routing messages in the network. The proposed scheme reduces the amount of messages retransmitted and for a network of 50 nodes, the improvement

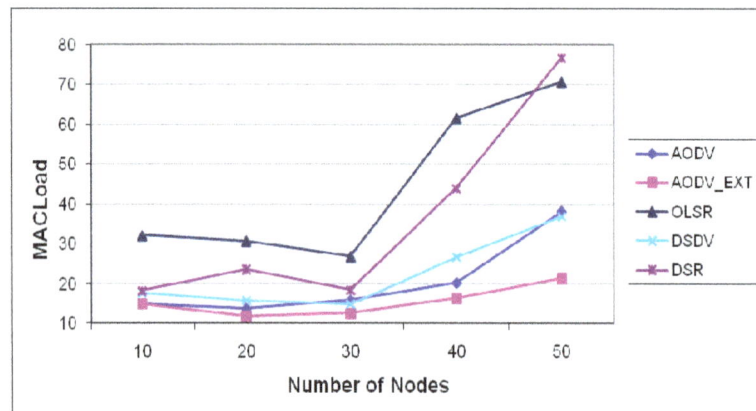

is by a factor of two over standard AODV.

Figure 3 MAC Loading against the number of nodes

Figure 4 shows the routing control message overheads. In the case of AODV_EXT, it is lower than that of standard AODV protocol. DSR and DSDV show better results due to the fact that they transmit and receive the less number of control messages. OLSR protocol has the worst performance of all the protocols and this degrades significantly as the number of nodes exceeds 20.

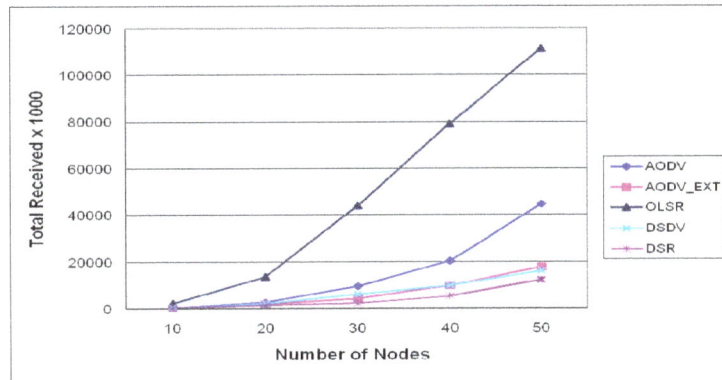

Figure 4 The Control Message Overhead

Table 2 provides details of the number of dropped packets for each protocol. DSDV drops the least number of packets with standard AODV dropping the most for a network of 50 nodes. However, compared to standard AODV, AODV_EXT performance is a factor 4 better.

Table 2: Number of dropped packets against network sizes

Number of Nodes	DSDV	DSR	AODV	AODV_EXT	OLSR
10	117	196	73	69	94
20	244	581	472	241	178
30	384	856	3011	766	473
40	911	4007	6537	1996	2978
50	1608	8880	17346	4056	7745

5. CONCLUSIONS

The assessment of four widely used protocols (AODV, DSDV, DSR and OLSR) has been presented in this paper. Their performances in different size networks and in mobile scenarios have been studied using simulations developed in Network Simulator 2 (NS2). AODV has been modified to use a probabilistic approach for transmitting route request messages. The modified version has been named AODV_EXT. Unlike in some probability based approaches where every node is assigned a fixed probability that does not guarantee full network coverage, the technique proposed in this paper combines concepts from maximum range node selection with node pruning to reduce redundant re-transmissions in route request but offer connectivity and better network coverage guarantees inherent in deterministic techniques.

The reduction in route request transmissions in a network using AODV_EXT has resulted in 3% energy efficiency savings, more than 60% reduction in the number of dropped packets because of reduced packet collision and increased data throughput. Moreover the results can be compared to that of Energy Reversed Ad-Hoc On-Demand Distance Vector (ER-AODV) routing protocol proposed by Khelifa and Maaza [27] which consumes up to 1.7 % more power than AODV_EXT. In addition, AODV_EXT improves the data throughput by more than 19% compared to the standard AODV and 10% more than ER-AODV. The results also show that proactive protocols, whilst they are more reliable in terms of connectivity, exhibit poor performance in large networks. Reactive protocols, on the other hand present better performances into large networks. Both classes of protocols perform poorly in large mobile networks due to large overheads associated with routing as the nodes move. A hybrid protocol such as AODV offer a compromise and the technique proposed in this paper to reduce redundant re-transmissions based on transmitting node neighbourhood density has produced very promising results when compared to standard protocols. This study has shown that fine tuning of protocols to suit specific applications or traffic scenarios to achieve optimum performance in ad-hoc networks will be essential.

References

[1]   R. Min, A. Chandrakasan, "Top Five Myths about the Energy Consumption of Wireless Communication", Mobile Computing and Communications Review, Vol. 7, No. 1, 2003.

[2]   R. Ramdhany, P. Grace, G. Coulson, D. Hutchison, "Dynamic Deployment and Reconfiguration of Ad-hoc Routing Protocols", Journal of Internet Services and Applications, Vol. 1, No. 2, pp. 135-152, 2010.

[3]   Y. Xu, J. Heidemann, D. Estrin, " Geography-informed Energy Conservation for Ad Hoc Routing", Proceedings of the Seventh Annual ACM/IEEE International Conference on Mobile Computig and Networking, Italy, 2001.

[4]   J. Rabaey, J. Ammer, J. L. da Silva Jr, D. Patel, "PicoRadio: Ad-Hoc Wireless Networking of Ubiquitous Low-Energy Sensor/Monitor Nodes", Proceeding WVLSI '00 Proceedings of the IEEE Computer Society Annual Workshop on VLSI (WVLSI'00), 2001.

[5]   V. Raghunathan, C. L. Pereira, M. B. Srivastava, R. K. Gupta, "Energy Aware Wireless Systems with Adaptive Power-Fidelity Tradeoffs", IEEE Transactions on Very Large Scale Integration (VLSI) Systems, Vol. 13, No. 2, Feb. 2005.

[6]   "Survey: Broadcasting in Wireless Sensor Network, Ad-hoc Network and Delay – Tolerant Network", Available: silentideas.com/wordpress/wp-content/.../broadcasting-jigar-report2.pdf

[7]   N. Karthikeyan, V. Palanisamy, K. Duraiswamy, "Optimum Density Based Model for Probabilistic Flooding Protocol in Mobile Ad-hoc Network", European Journal of Scientific Research, Vol. 39, No. 4, pp. 577-588, 2010.

[8]   N. Karthikeyan, V. Palanisamy, K. Duraiswamy, "Reducing Broadcast Overhead Using Clustering Based Broadcast Mechanism in Mobile Ad-hoc Network", Journal of Computer Science, pp. 548-556, Vol. 5, No. 8, 2009.

[9]   R. Min, A. Chandrakasan, "Top Five Myths about the Energy Consumption of Wireless Communication", MIT, http://www.sigmobile.org/mobicom/2002/posters/rex-min.pdf, accessed 3 Nov. 2010.

[10] J. Wu, F. Dai, "Efficient Broadcasting with Guaranteed Coverage in Mobile Ad-hoc Networks", IEEE Transactions on Mobile Computing, pp. 259-270, Vol. 4, No. 3, June 2005.

[11] H. Abusaimeh, S. Yang, "Reducing the transmission and reception powers in the AODV", Proceedings of International Conference on Networking, Sensing and Control, pp. 60-65, March 2009.

[12] A. Jamal, "Probabilistic route discovery for Wireless Mobile Ad-hoc Networks (MANETs), PhD thesis, University of Glasgow, 2009.

[13] J. C. Requena, T. Vadar, R. Kantola, N. Beijar, "AODV-OLSR scalable ad hoc routing proposal", Symposium on Wireless Pervasive Computing, 2006

[14] L. Junhai, X. Liu, Y. Danxia, " Research on multicast routing protocols for mobile ad-hoc networks" Science Direct, pp. 988-997, Vol. 52 Issue 5, April 2008

[15] M. Abolhasan, T. Wysocki, E. Dutkiewicz, " A review of routing protocols for mobile ad-hoc networks ", Science Direct, pp. 1-22, Vol, 2, Issue 1, January 2004.

[16] H. Minemo, K. Soga, T. Takenaka, Y. Terashima, T. Mizuno, "Integrated protocol for optimized link state routing and localization: OLSR-L", CISIS '10 Proceedings of the 2010 International Conference on Complex, Intelligent and Software Intensive Systems, 2010

[17] A. H. Abd Rahman, Z. A. Zukarnain, " Performance Comparison of AODV, DSDV and I-DSDV Routing Protocols in Mobile Ad-hoc Networks ", European Journal of Scientific Research, pp. 556-576, Vol.31, No.4, June 2009.

[18] D. Maltz, Y. Hu, "The Dynamic Source Routing Protocol for Mobile Ad-hoc Networks", Internet Draft, Available:http://www.ietf.org/internet-drafts/draft-ietf-manet-dsr-10.txt, July 2004.

[19] C. Perkins and E. Royer, "Ad-hoc On-demand Distance Vector (AODV) Routing", Internet Draft, MANET working group, draft-ietf-manet-aodv-05.txt, March 2000.

[20] F. Rango D., M. Fotino, S. Marano, "EE-OLSR: Energy Efficient OLSR Routing Protocol for Mobile Ad-Hoc Networks", Military Communications Conference, IEEE, pp. 1-7, Nov. 2008.

[21] A. Qayyum, L. Viennot, A. Laouiti, "Multipoint Relaying for Flooding Broadcast Messages in Mobile Wireless Networks," 35th Annual Hawaii International Conference on System Sciences (HICSS'02), pp.3866-3875, Volume 9, Jan. 2002.

[22] "Ns-2 network simulator," http://www.isi.edu/nsnam/ns/, 1998.

[23] "CMU Monarch extensions to ns2," http://www.monarch.cs.cmu.edu/cmu-ns.html, 1999.

[24] Marc Greis' Tutorial for the UCB/LBNL/VINT Network Simulator "ns", "http://web.uct.ac.za/depts/commnetwork/tutorial_ns_full.pdf", 2006.

[25] "Installing OLSR on ns2",http://masimum.inf.um.es/um-olsr/html, 2005.

[26] A. Rahman, S. Islam, A. Talevski, "Performance Measurement of various Routing Protocol in Ad-Hoc Network ", IMECS, Vol. 1, March 2009, pp. 321-323.

[27] S. Khelifa, M. Maaza, "An Energy Multi-path AODV routing protocol in ad hoc mobile networks", 5th International Symposium on I/V Communications and Mobile Network (ISVC), Rabat,Morocco, 2010.

14

Performance Evaluation of Routing Protocols in MANETS

Humaira Nishat[1], Sake Pothalaiah[2] and Dr. D.Srinivasa Rao[3]

[1]Department of Electronics and Communication Engineering, CVR College of Engineering, Hyderabad, India.
huma_nisha@yahoo.com
[2]Department of Electronics and Communication Engineering, MRP UGC, Jawaharlal Nehru Technological University, Hyderabad, India.
pawan.s14@gmail.com
[3]Department of Electronics and Communication Engineering, Jawaharlal Nehru Technological University, Hyderabad, India.
dsraoece@jntuh.ac.in

ABSTRACT

In wireless ad hoc networks mobile stations or nodes are free to move around. The transmission range of the nodes is fixed in mobile ad hoc networks (MANETs) whereas the network topology changes in a different fashion. Due to dynamic nature of network topology some of the network links are destroyed while some new links are established. The routing protocols developed for wired networks cannot be used efficiently for wireless networks. For wireless ad hoc networks there are a few new routing protocols suitable for the dynamically changing ad hoc wireless environment. In this paper we compare the performance of two on-demand routing protocols (AODV and DSR) in terms of QoS parameters such as throughput, minimum, maximum & average delay and packet delivery ratio. We performed extensive simulations using NS-2 simulator using both conventional TCP and TCP Vegas traffic sources.

Keywords

AODV, DSR, MANET routing protocols, TCP & TCP Vegas.

1. INTRODUCTION

Mobile networks are classified as infrastructure networks and Mobile ad hoc networks (MANETs) [1], [2]. In infrastructure mobile network, nodes have basestations or wired access points within their transmission range. In contrast, MANETs [1] are autonomous self-organized networks without support of infrastructure. Mobile stations in MANETs are free to move around. Because of the fixed transmission range of mobile terminals, the network topology changes dynamically resulting in network establishment and breaking of some existing network links.

For wired networks, routing protocols were developed with the assumption that the topology is static. Therefore such routing protocols may not serve efficiently in case of wireless ad hoc networks. Thus new routing protocols are developed for the dynamically changing [8] ad hoc wireless environment. The routing protocols of wireless ad hoc networks fall into two category (1) Table-driven and (2) On-demand [9].

1.Table driven routing protocol maintain consistent, up-to-date routing information from each node to all other nodes of the network. Each network node therefore maintains one or more routing table which stores the routes to all the other network nodes. When changes in topology

[8] occurs, the related information is sent to all network nodes in order to provide up-to-date routing information. Table driven routing protocol have the disadvantage of increased signaling traffic and power consumption as the routing information is disseminated to all the network nodes.

2.On-demand routing protocol follow a different approach. A route is established only when there is a need to for a network connection. When a source node X needs a connection to a destination Y, it invokes a routing discovery protocol to find a route connecting it to Y. Once the route establishment is done, nodes X & Y and all the intermediate nodes store the information regarding the route from X to Y in their routing tables. The route is maintained until the destination is unreachable or the route is no longer needed. On-demand routing protocols have lower power consumption and less control signaling however, it has long end-to-end connection delay as the connection is established only upon the generation of a network connection required. In wireless ad hoc networks routing protocols are developed assuming that all stations have identical capabilities and employ the capability to perform routing related tasks such as route discovery/establishment and route maintenance in the network.

Several performance evaluation of routing protocols in MANETs have been performed using CBR traffic. Biradar , S.R. et al [10] have analyzed the AODV and DSR protocol using Group mobility model and CBR traffic sources. According to [10] DSR performs better in high mobility and DSR gives better average delay. Rathy R.K. et al [11] compared AODV and DSR routing protocols under random way point mobility model with TCP and CBR traffic sources. According to [11] AODV outperforms DSR in high mobility and/or high load situations. Harminder S.B. et al[12] investigated the performance of AODV and DSR routing protocol under group mobility models. According to [12] DSR gives better results in TCP traffic and under restricted bandwidth condition.

In this paper we have investigated the performance of on-demand routing protocols such as Ad hoc on demand distance vector (AODV)[3] and Dynamic source routing (DSR)[4] routing protocols in the scenario of Random Mobility Model using both conventional TCP and TCP Vegas traffic sources. The objective of the work is to understand the working mechanisms and to investigate which routing protocol gives better performance when TCP and TCP Vegas are used as the traffic source.

The rest of the paper is organized as follows. Section 2 gives a brief introduction of both AODV and DSR routing protocols. Section 3 tells about the simulation setup. Section 4 gives the results and performance comparison of the routing protocols. Finally, section 5 concludes the paper.

2. DESCRIPTION OF ROUTING PROTOCOLS

2.1 Ad hoc on demand distance vector (AODV)

Ad hoc on demand distance vector (AODV) [3] routing protocol creates routes on-demand. In AODV, a route is created only when requested by a network connection and information regarding this route is stored only in the routing tables of those nodes that are present in the path of the route.

The procedure of route establishment is as follows. Assume that node X wants to set up a connection with node Y. Node X initiates a path discovery process in an effort to establish a route to node Y by broadcasting a Route Request (RREQ) packet to its immediate neighbors. Each RREQ packet is identified through a combination of the transmitting node's IP address and a broadcast ID. The latter is used to identify different RREQ broadcasts by the same node and is incremented for each RREQ broadcast. Furthermore, each RREQ packet carries a sequence number which allows intermediate nodes to reply to route requests only with up-to-date route information. Upon reception of an RREQ packet by a node, the information is forwarded to the immediate neighbors of the node and the procedure continues until the RREQ is received either by node Y or by a node that has recently established a route to node Y. If subsequent copies of the same RREQ are received by a node, these are discarded.

When a node forwards a RREQ packet to its neighbors, it records in its routing table the address of the neighbor node where the first copy of the RREQ was received. This helps the nodes to establish a reverse path, which will be used to carry the response to the RREQ. AODV supports only the use of symmetric links. A timer starts running when the route is not used. If the timer exceeds the value of the 'lifetime', then the route entry is deleted.

Routes may change due to the movement of a node within the path of the route. In such a case, the upstream neighbor of this node generates a 'link failure notification message' which notifies about the deletion of the part of the route and forwards this to its upstream neighbor. The procedure continues until the source node is notified about the deletion of the route part caused by the movement of the node. Upon reception of the 'link failure notification message' the source node can initiate discovery of a route to the destination node.

2.2 Dynamic Source Routing (DSR)

Dynamic Source Routing (DSR) [4] uses source routing rather than hop-by-hop routing. Thus, in DSR every packet to be routed carries in its header the ordered list of network nodes that constitute the route over which the packet is to be relayed. Thus, intermediate nodes do not need to maintain routing information as the contents of the packet itself are sufficient to route the packet. This fact eliminates the need for the periodic route advertisement and neighbor detection packets that are employed in other protocols. The overhead in DSR is large as each packet must contain the whole sequence of nodes comprising the route.

DSR comprise the processes of route discovery and route maintenance. A source node wishing to set up a connection to another node initiates the route discovery process by broadcasting a RREQ packet. This packet is received by neighboring nodes which in turn forward it to their own neighbors. A node forward a RREQ message only if it has not yet been seen by this node and if the nodes address is not part of route. The RREQ packet initiates a route reply packet (RREP) upon reception of the RREQ packet either by the destination node or by an intermediate node that knows a route to the destination. Upon arrival of the RREQ message either to the destination or to an intermediate node that knows a route to the destination, the packet contains the sequence of nodes that constitute the route. This information is piggybacked on to the RREP message and consequently made available at the source node. DSR supports both symmetric and asymmetric links. Thus, the RREP message can be either carried over the same path with original RREQ, or the destination node might initiate its own route discovery towards the source node and piggyback the RREP message in its RREQ.

In order to limit the overhead of these control messages, each node maintains a cache comprising routes that were either used by these nodes or overheard. As a result of route request by a certain node, all the possible routes that are learned are stored in the cache. Thus, a RREQ process may result in a number of routes being stored in the source node's cache.

Route maintenance is initiated by the source node upon detection of a change in network topology that prevents its packet from reaching the destination node. In such a case the source node can either attempt to use alternative routes to the destination node or reinitiate route discovery. Storing in the cache of alternative routes means that route discovery can be avoided when alternative routes for the broken one exist in the cache. Therefore route recovery in DSR can be faster than   any other on-demand routing protocols.

Since route maintenance is initiated only upon link failure, DSR does not make use of periodic transmissions of routing information, resulting in less control signaling overhead and less power consumption at the mobile nodes.

3. SIMULATION SETUP

We have used network simulator version 2.34 for the evaluation of our work. The NS-2 simulator software was developed at the University of California at Berkeley and the Virtual Inter Network testbed (VINT) project fall 1997[5],[6]. We have used Ubuntu 9.04 Linux environment. Our simulation setup [9] is a network with randomly placed nodes within an area of 1315m * 572m. We have chosen a wireless channel with a two-way ground propagation model with a radio propagation model of 250m and interference range of 550m. The parameters used for carrying out simulation are summarized in the table1.

Table 1. Simulation Parameters

Parameter	Value
Routing Protocols	AODV,DSR
MAC Layer	802.11
Terrain Size	1315m*572m
No. of Nodes	25
Mobility Model	Random Mobility Model
Packet Size	1500B
Bandwidth	11MB
Frequency	2.472GHz
Antenna Type	Omni antenna
Propagation Model	2-Ray ground
Speed	0-5-10-15-20-25m/s
Simulation Time	100s
Traffic Source	TCP, TCP Vegas
Application Layer	FTP

The node's speed is varied from 0 to 25m/s generated by uniform distribution. The simulation execution time is 100s. We have simulated the scenario with both the conventional TCP and TCP Vegas traffic sources. The aim of our simulation is to evaluate the performance differences of the two on-demand routing protocols and compare it with both TCP and TCP Vegas.

3.1 Performance metrics

Manet routing protocols can be evaluated by a number of quantitative metrics described by RFC2501 [7]. We have used the following metrics for evaluating the performance of the two routing protocols (AODV & DSR).

3.1.1 Packet Delivery Fraction

It is the ratio of the number of packets received by the destination to the number of data packets generated by the source.

3.1.2 Minimum Delay

It is defined as the minimum time taken for a data packet to be transmitted across a MANET from source to destination.

3.1.3 Maximum Delay

It is defined as the maximum time taken for a data packet to be transmitted across a MANET from a source to destination.

3.1.4 Average end-to-end delay

It is defined as the average time taken by the data packets to propagate from source to destination across a MANET. This includes all possible delays caused by buffering during routing discovery latency, queuing at the interface queue, and retransmission delays at the MAC, propagation and transfer times.

3.1.4 Throughput

It is the rate of successfully transmitted data packets per second in the network during the simulation.

4. SIMULATION RESULTS

Here we present a comparative analysis of the performance metrics of both the on-demand routing protocols AODV and DSR with both TCP and TCP Vegas traffic sources for different node speeds 5,10,15,20 & 25m/s.

4.1 Packet Delivery Fraction

In case of TCP traffic source at low node velocity i.e., from 0 to 15m/s DSR performs better than AODV. But as the speed increases to 20m/s both DSR and AODV performs equally under all assumed load condition. With TCP Vegas, DSR gives more PDF than AODV at both low as well as high node velocities (Fig1). At low velocities AODV is comparable to DSR but the ratio decreases as the speed of node increases. Thus we conclude that AODV with TCP Vegas is comparable to DSR at low velocities of node but at high node velocities DSR performs better.

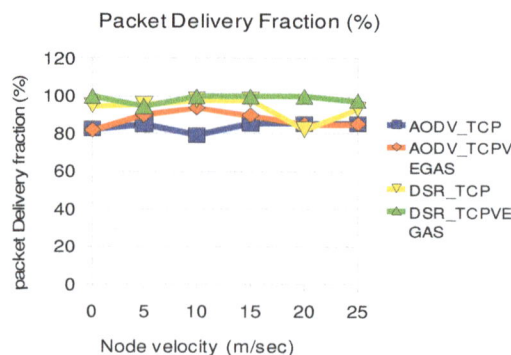

Figure 1. Packet Delivery Fraction

4.2 Minimum Delay

In case of TCP traffic source, minimum end-to-end delay of AODV is better than DSR. As the velocity of node increases to 25m/s DSR has maximum delay. Thus AODV outperforms DSR. Comparing AODV with conventional TCP and TCP Vegas, at low speeds from 0 to 15m/s both traffic sources generates equal delay but as the speed increase to 25m/s AODV with TCP Vegas gives the minimum amount of delay which is 14msec.

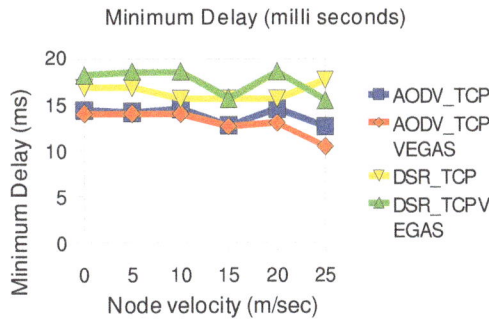

Figure2. Minimum Delay

4.3 Maximum Delay

With TCP traffic source AODV gives almost constant delay at all the speeds whereas delay of DSR increases almost linearly as the speed of node increases. DSR gives the maximum delay of 4sec at 25m/s. With TCP Vegas traffic source both AODV and DSR has equal delay at 0m/s but as the node velocity increases to 5m/s, delay of DSR increases abruptly to a high value and thereby remains constant till 25m/s. Comparing conventional TCP and TCP Vegas traffic sources DSR gives almost equal delay till 20m/s but at 25m/s DSR with TCP gives a maximum delay of $4*10^3$ m sec. Thus AODV with TCP Vegas gives less delay thereby outperforming DSR.

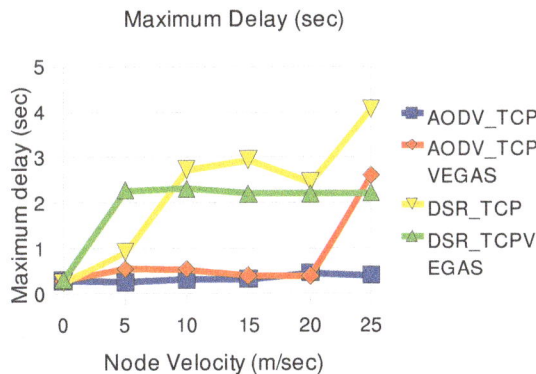

Figure3. Maximum Delay

4.4 Average Delay

In case of conventional TCP , AODV gives almost constant and least delay at all the node velocities whereas delay of DSR increases with the node velocity. In case of TCP Vegas also, AODV gives less delay than DSR.Thus the average end-to-end delay is least for AODV routing protocol with TCP Vegas traffic source.

Average Delay (sec)

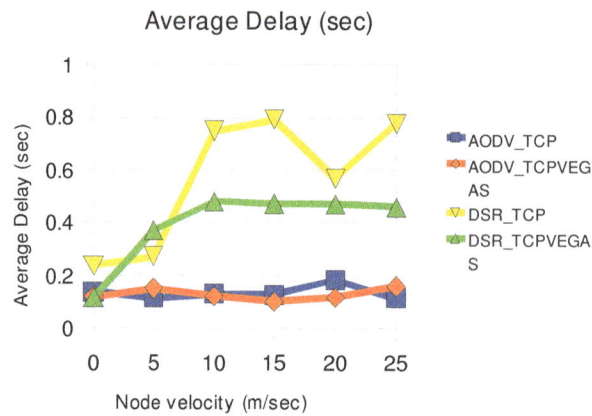

Figure 4.Average Delay

4.5 Throughput

In case of TCP traffic source, at 0m/s both DSR and AODV gives equal and maximum throughput. At 10m/s AODV gives less throughput than DSR but as the speed increases AODV outperforms DSR. With TCP Vegas traffic source, at 0m/s DSR gives more throughput than AODV but as the speed increases throughput of DSR decreases. Thus, AODV performs better than DSR as the speed increases. At low node velocities, AODV with both TCP & TCP Vegas performs equally. But at higher velocities, AODV with conventional TCP gives better throughput performance than with TCP Vegas.

Throughput (Kbps)

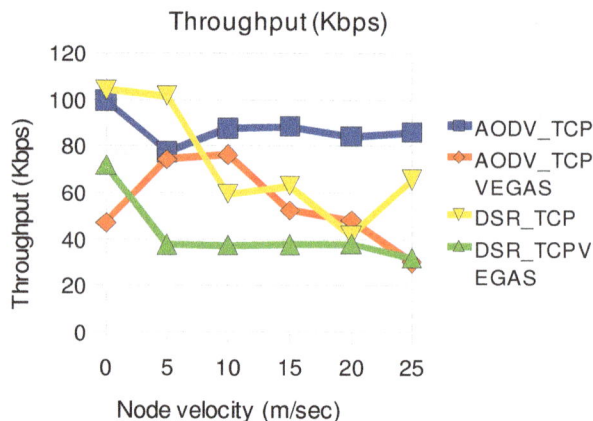

Figure5. Throughput

5. CONCLUSIONS

We have evaluated the two routing protocols AODV and DSR using both TCP & TCP Vegas traffic sources. Based on the results we conclude that, both AODV and DSR gives almost same packet delivery fraction at low node velocities but as the velocity of the node increases DSR gives better PDF with TCP Vegas. Delay is maximum for DSR and minimum for AODV with TCP Vegas. Average end-to-end delay of AODV is less than DSR. Throughput of AODV is better than that of DSR. Thus, AODV with TCP Vegas traffic source outperforms DSR.

In this paper the two routing protocols AODV & DSR are analyzed and their performances have been evaluated with respect to five performance metrics using the two traffic sources TCP & TCP Vegas. This paper can be enhanced by analyzing other MANET routing protocols with different traffic sources.

6. REFERENCES

[1] M.S. Carson, S. Batsell and J. Macker,   "Architecture consideration for Mobile Mesh Networking," Proceedings of the IEEE Military Communications Conference(MILCOM), vol.1, pp 225-229, 21-24 oct.1996.

[2] C.K. Toh "Ad Hoc Mobile Wireless Networks   Protocols and Systems", First Edition, Prentice Hall Inc, USA 2002.

[3] C.Perkins and E.Royer, "Ad Hoc On Demand Distance Vector Routing," Proceedings of the 2nd IEEE workshop on Mobile Computing Systems and Applications(WMCSA 1999) pp.99-100, Feb-1999.

[4] D.B.Johnson, D.A.Maltz and J.Broch, "DSR: The Dynamic Source Routing Protocol for Multi-Hop Wireless Ad Hoc Networks", Ad Hoc Networking, pp 139-172, 2001.

[5] UCB/LBNL/VINT Network Simulator, http://wwwmash.cs.berkeley.edu/ns/referred on March 2010.

[6] "The Network Simulator- ns-2", available at http://www.isi.edu/nsnam/ns/referred on march 2010.

[7] S.Corson and J.Macker, "Routing Protocol performance Issues and Evalaution Considerations", RFC2501, IETF Network Working Group, January 1999.

[8] N.H.Vaidya, "Mobile Ad Hoc Networks Routing, MAC and transport Issues", Proceedings of the IEEE International Conference on Computer Communication INFOCOM,2004.

[9] L.Layuan, Y.Peiyan and L.Chunlin, "Performance Evaluation and Simulations of Routing Protocols in Ad Hoc Networks" Computer Communications, vol.30, pp.1890-1998, 2007.

[10] S.R.Biradar, Hiren H.D.Sharma, Kalpana Sharma and Subir Kumar Sarkar, "Performance Comparison of Reactive Routing Protocols of MANETs Using Group Mobility Model", IEEE International Conference on Signal Processing Systems, pages 192-195, 2009.

[11] Suresh Kumar, R.K.Rathy and Diwakar Pandey, "Traffic Pattern Based Performance Comparison of Two Routing Protocols for Ad hoc Networks Using NS2", 2nd IEEE International Conference on Computer Science and Information Technology, 2009.

[12] Harminder S.Bindra, Sunil K.Maakar and A.L.Sangal, "Performance Evaluation of Two Reactive Routing Protocols of MANET using Group Mobility Model", International Journal of Computer Science Issues, Vol.7, Issue 3, No.10, May 2010.

[13] Yasser Kamal Hassan, Mohammed Hashim Abd El-Aziz and Ahmed Safwat Abd El-Radi, "Performance Evaluation of Mobility Speed over MANET Routing Protocols", International Journal of Network Security, Vol.11, NO.3, pp.128-138, Nov 2010.

[14] X.Hong, M.Gerla, G.Pei and C.C.Chiang, "A Group Mobility Model for Ad Hoc Wireless Networks", in ACM/IEEE MSWiM, August1999.

[15] M. Li, L. Zhang, V. O. K. Li, X. Shan and Y. Ren, "An Energy-Aware Multipath Routing Protocol for Mobile Ad Hoc Networks," Proceedings of the ACM SIGCOMM Asia, April 2005.

[16] N. T. Javan and M. Dehghan, "Reducing End-to-End Delay in Multi-path Routing Algorithms for Mobile Ad hoc Networks," Proceedings of the International Conference on Mobile Ad hoc and Sensor Networks (MSN 2007), Lecture Notes in Computer Science (LNCS) 4864, pp. 703 – 712, December 2007.

[17] S. Vijay, S. C. Sharma, V. Gupta and S. Kumar, "CZM-DSR: A New Cluster/Zone Disjoint Multi-Path Routing Algorithm for Mobile Ad Hoc Networks," Proceedings of the IEEE International Advanced Computing Conference (IACC), Patiala, India, pp. 480 – 485, March 2009.

[18] I. S. Ibrahim, A. Etorban and P. J. B. King, "Multi-path Distance Vector Zone Routing Protocol for Mobile Ad hoc Networks: MDVZRP," Proceedings of the 9th Annual Postgraduate Symposium on The Convergence of Telecommunications, Networking and Broadcasting (PGET), June 2008.

[19] J. Zhang, C. K. Jeong, G. Y. Lee and H. J. Kim, "Cluster-based Multi-path Routing Algorithm for Multi-hop Wireless Network," International Journal of Future Generation Communication and Networking, Vol. 1, No. 1, pp. 67 – 74, December 2008.

[20] D. Shin and D. Kim, "3DMRP: 3-Directional Zone-Disjoint Multi-path Routing Protocol," IEICE Transactions on Information and Systems, Vol. E92-D, No. 4, pp. 620-629, April 2009.

[21] G. G. Md. N. Ali, R. Chakraborty, Md. S. Alam and E. Chain, "An Efficient Approach for Generalized Load Balancing in Multi-path Packet Switched Networks," AIRCC International Journal of Computer Networks and Communications, Vol. 2, No. 2, pp. 142-153, March 2010.

[22] N. Cooper and N. Meghanathan, "Impact of Mobility Models on Multi-path Routing in Mobile Ad hoc Networks," AIRCC International Journal of Computer Networks and Communications, Vol. 2, No. 1, pp. 185-174, January 2010.

[23] Natarajan Meghanathan, "Performance comparison of link, node and zone disjoint multi-path routing strategies and minimum hop single path routing for mobile ad hoc networks", AIRCC International Journal of Wireless & Mobile Networks (IJWMN) Vol.2, No.4, November 2010

[24] R. Manoharan and E. Ilavarasan "Impact of Mobility on the Performance of Multicast Routing Protocols in Manets", AIRCC International Journal of Wireless & Mobile Networks (IJWMN) Vol.2, No.2, May2010.

15

IMPROVING ENERGY EFFICIENCY IN MANETS BY MULTI-PATH ROUTING

Hassanali Nasehi[1], Nastooh Taheri Javan[1], Amir Bagheri Aghababa[1] and Yasna Ghanbari Birgani[2]

[1]Department of Engineering, East Tehran Branch, Islamic Azad University (IAU)
Tehran, Iran
[2]Department of Industrial Engineering, Tarbiat Modares University (TMU)
Tehran, Iran
hanasehi@yahoo.com, nastooh@aut.ac.ir, amir_baqeri_aqababa@yahoo.com,
yasna.ghanbari@modares.ac.ir

ABSTRACT

Some multi-path routing algorithm in MANET, simultaneously send information to the destination through several directions to reduce end-to-end delay. In all these algorithms, the sent traffic through a path affects the adjacent path and unintentionally increases the delay due to the use of adjacent paths. Because, there are repetitive competitions among neighboring nodes, in order to obtain the joint channel in adjacent paths. The represented algorithm in this study tries to discover the distinct paths between source and destination nodes with using Omni directional antennas, to send information through these simultaneously. For this purpose, the number of active neighbors is counted in each direction with using a strategy. These criterions are effectively used to select routes. Proposed algorithm is based on AODV routing algorithm, and in the end it is compared with AOMDV, AODVM, and IZM-DSR algorithms which are multi-path routing algorithms based on AODV and DSR. Simulation results show that using the proposed algorithm creates a significant improvement in energy efficiency and reducing end-to-end delay.

KEYWORDS

MANET, Multi-path Routing, Energy Efficiency.

1. INTRODUCTION

MANETs are such networks in which there are no infrastructures [1]. Therefore, all network functions such as routing and data transfer are performed by the nodes themselves and by cooperating with each other. These types of networks, the routing process has its special difficulty and complexity due to the high mobility of nodes and dynamic network topology. From the beginning of the mobile networks, many routing algorithms for these networks have been suggested. Some of those algorithms are AODV and DSR that have more popularity and acceptability than other algorithms [2]. Both of these algorithms relates to the class of on-demand routing algorithms. In the on- demand routing algorithm, the path discovery process begins when a node has a packet to send without any valid path to its specific destinations.

Among the mass network routing algorithms in mobile ad hoc networks, multi-path algorithms have found their place [3]. In these algorithms, it tends to discover several paths between source and destination rather than finding a path between them .The main advantage of this idea is that the time consuming process is executed less times to discover paths. And when one of the routes faces with failure, it is possible to use one another discovered path quickly. Among the multi-path routing algorithms in MANET, Some of them after discovering some paths between origin and destination start sending data through several paths simultaneously, in order to reduce latency and increase end to end bandwidth.

One of the most important issues in any kind of multi-path routing algorithms is how to select multiple paths. These algorithms, try to use more discrete paths as possible to increase the reliability from one side and reduce shared resources, increase bandwidth and reduce latency from another side. In fact, these algorithms, in their best status, prefer to use node-disjoint paths. In node disjoint paths, there is no node between two joint paths, and, therefore paths are completely independent and they don't have any shared resources. Thus, paths' bandwidth doesn't have anything in common, and by the deterioration of a node or a connection, at most one of the paths will be lost.

Up to now, by discovering and choosing node-disjoint paths, everything seems to be good. However, by focusing on the essence and the structure of MANET, a new problem comes up. As we know, in MANET, there are inherent problems such as Exposed Terminal Problem and Hidden Terminal Problem. To resolve these problems, CSMA/CA protocol has been proposed [4] which is used for accessing the channels in 802.11 standard. In this protocol, due to exchanging the RTS and CTS messages between nodes, some nodes are forced to avoid sending information. This fact will increase the end-to-end delay [5].

For instance, figure 1 shows a hypothetical network in which ten nodes is exhibited. Also, the radio range of each node is specified and the dash lines indicate direct connection between two nodes. In other words, the dash line between two specific nodes means that two nodes are in radio range of each other. In this network, between nodes S and D, there are two distinct paths S-I1-I2-I3-I4-D and S-I5-I6-I7-I8-D (upper and lower paths) in which communication and data transmission through one path are not completely independent from another path. In this state, the end-to-end delay of each path depends on the traffic of the other path. This is because of the exchange of RTS and CTS messages among nodes of network for avoiding collisions and salvation of Exposed Terminal Problem and Hidden Terminal Problem. As a result, some terminals of a path should postpone their sending process in order to receive CTS from a node in the opposite path, for instance.

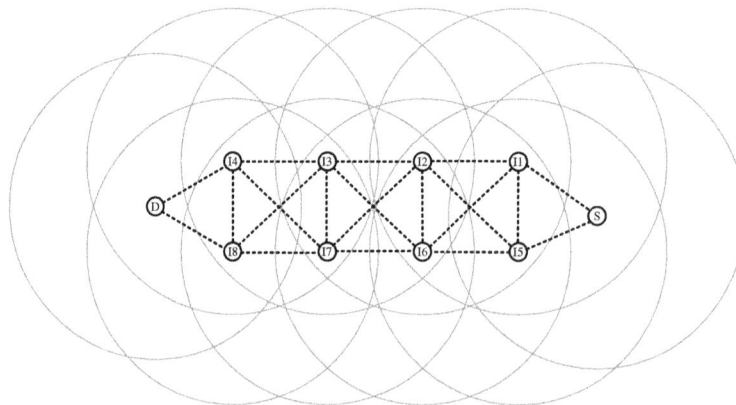

Figure 1. Node-Disjoint Paths.[5]

To resolve this problem, several methods have been proposed in which directional antennas are used [6]. Also we can use zone-disjoint paths instead of node-disjoint paths [5]. Two routes with no pair of neighbor nodes are called zone-disjoint in terminology.

In this paper, we have tried to resolve this problem somehow, by modifying the AODV routing algorithm and using omni-directional antennas. In our previous work [7] we focused on end-to-end delay, but in present work we have tried to improve energy efficiency in MANET.

This article contains following parts. In the second part of this article, we have analyzed earlier works around the same topic. In Section 3, we go through the details of the proposed algorithm. Also, in section 4, we explain the simulation results of our approach. Finally, in Section 5, a brief conclusion of the work is represented.

2. RELATED WORKS

A lot of approaches have been made in multi-path routing in MANET [8, 9, 10]. Moreover, single-path algorithms have been applied in such algorithms due to their good performance. One of these single-path algorithms is AODV. Also, AOMDV algorithm has been suggested based on AODV [11]. AOMDV algorithm has become popular because of its simplicity. This algorithm tries to discover distinct disjoint-paths between origin and destination. In this state, source creates a path request message and broadcasts it to all of its neighbors. Unlike single-path algorithm in which analogous messages are deleted, this algorithm doesn't delete repeated messages. Since the distinct paths should be discovered, each of these source neighbors will be the beginner of a new feasible path. Source neighboring nodes rebuild request package and re-broadcast it by entering their address in RREQ. Each intermediate node receives and analyzes this packet. If middle nodes have not responded to such request, they receive the packet and rebroadcast it with new number of Hop Counts. But if such a packet was previously replied, and if the following conditions are satisfied, the packet will be accepted and inserted in the node table.

1. If the packet is received a from different neighbor node
2. The packet is received from this source neighbor node which we have not had any path from it so far. (This packet is a presenter of a new path).
3. If the number of hop counts is less than the existing values along the way.

The destination node, upon receiving the path request packet, sends a desired number of path reply message to the request packets for this request. Therefore, paths can be elicited at the source point.

AODVM [12] is also based on AODV, and can find only node disjoint paths. The scheme proposed in AODVM takes advantage of reliable nodes along with multipath routing to construct a reliable path between two nodes. Packet salvaging can be used in multipath routing to provide improved fault tolerance.

Zone disjoint multi-paths are implemented with using directional antennas [6]. In this method, nodes are constantly inserting neighbor's information including power and neighbor signal angle, in a table. Among the existing paths, node disjoint paths are selected, and then zone distinction of the paths will be determined by each node's tables. This work will be constantly repeated to always ensure finding zone disjoint paths. One of the disadvantages of this method is that the directional antennas must be used. However, multi-directional antennas are used in most ad-hoc networks and directional antennas are not available .

In [13], the idea of discovering zone disjoint paths in Source Routing algorithms with using Omni directional antennas has been explored, and in [14] this idea has been implemented and applied in DSR base routing algorithm. In this algorithm, after discovering the zone disjoint paths from source to destination, it has been tried to send the information to destination via discovered paths simultaneously. Also in this algorithm, the idea of counting the number of active neighbors is performed in order to discover zone disjoint paths. In our previous work [7] we implemented this method in AODV.

3. SUGGESTED ALGORITHM

The suggested algorithm has been designed and implemented based on AODV algorithm. The AODV algorithm is considered to be in the class of on-demand routing algorithms in which routing process takes place hop by hop. In this way, each node has a path table in which received packet's information are saved.

As it is mentioned in the introduction, the proposed algorithm tries to discover zone disjoint paths between source and destination in order to send information simultaneously. If there is possibly no neighboring between two nodes in two distinct paths, the paths are called area distinct. Briefly, the proposed algorithm counts the number of active neighbors for each path, and finally it chooses some paths for sending information in which each node has lower number of active neighbors all together. Here, active neighbors of a node are defined as nodes that have previously received the RREQ. There is this possibility that source and destination choose another path with nodes to exchange information; thus, information exchanging depends on this path. In fact, these two nodes are on two disjoint but adjacent paths.

3.1. Necessary alterations in the AODV algorithm

In most of the implementations of the AODV algorithm, the middle nodes maintain a Route Cache table in which they put paths that have been discovered during the course of the path discovery. Thus, if an middle node receives a packet containing path request from a predetermined source, it will return a path reply packet to the source. However, in the proposed algorithm, middle nodes don't need to maintain Route Cache tables. Therefore, more path request packets will reach destination. In fact, all the path request packets move from source to destination.

In addition, in the suggested algorithm, each node must put the received RREQ specifications in the table which is called RREQ_Seen, in order to respond to the neighbor queries properly. Also, to count the active neighbors in each path, in the RREQ_Seen table, each node has a field named "the number of active neighbors after sending RREQ" that this field is briefly called After_A_N_C in this article. Also, the ActiveNeighborCount field name is added to the headers of RREQ and RREP to make next nodes of the path aware of the number of neighboring nodes in traversed nodes. Finally, two new packet as RREQ_Query and RREQ_Query_Reply are added to the path discovery process to perform the query process. More thoroughly, the query initiator node places current RREQ profile information into RREQ_Querypacket and sends it to its neighbors. If nodes themselves are the answer of the auery process they turn back a RREQ_Query_Reply to the initiator.

3.2. The suggested algorithm's procedures

When a node is about to send data to a specific destination and it does not find a valid path to its destination, the node runs the path discovery process by producing and sending RREQ packet to its neighbors. In this RREQ packet, the initial value of zero will be assigned to ActiveNeighborCount field. Therefore, source neighbor nodes receive RREQ packet, set their names as the founders of one of the paths and reversely put the path specifications into the path table. But before resending the RREQ packet, the neighbor nodes request query path from their neighbors. In fact, they ask all their neighbors: " Have you seen a RREQ with this specification?" Then they increase the value of ActiveNeighborCount in RREQ packet for those neighbors which have a positive answer to this question. For this query, nodes use some packets with titles of RREQ_Query and RREQ_Query_Reply. Actually, the query node sends the RREQ_Query packet to its neighbors and after specific time period (which is calculated by a clock) waits for neighbors' responses to the question. On the other hand, all neighboring nodes are required to search the specification of RREQ in RREQ_Seen table after receiving the query packet. If neighboring nodes have already observed this RREQ, they reply a positive response to the query node. The response to query is performed by the production and transmission of RREQ_Query_Reply packet. Finally, after the time expiration, the node that has created the query broadcasts the RREQ packet to continue the discovery process.

Once again, we analyze the behavior of queried node. Since the repeated RREQ packets aren't removed in discovery of multiple paths, it is possible for a node to receive the RREQ packet for the second time. Therefore, it initiates the query process to discover new possible neighbors for the second time to. But obviously, only new neighbors need to consider this query important and old neighbors shouldn't answer to this repeated query. Thus, those nodes that receive the query packet keep the address and details of the query node and the queried RREQ packet in Query_Seen table. If a node receives a query packet for the first time, it sends a RREQ_Query_Reply packet to inform query node after recording a query's specifications. But if this query has already been received from the same node, it is not noticed.

Now, we focus on another aspect of query node. It is possible that a node to be queried by a new node after receiving a RREQ, performing query, updating the ActiveNeighborCount field in RREQ packet, and finally sending the RREQ (Further, this scenario is described in an example.) In fact this new neighbor is not considered in computing of the node. To resolve this problem, a field as After_A_N_C is added to RREQ_Seen table of each node. Now, after rejecting RREQ packet, if the node has been queried about the RREQ before the RREP arrivals, it adds one unite to the value of after_A_N_C field for each query. Noticeably, this node is still answering the query positively so that the query node can calculate accurately. After these measurements, when the RREP is sent from the source to the destination, in the middle of the path, each node adds the value of After_A_N_C in its RREQ_Seen table to ActiveNeighborCount field in RREP packet.

Thus, when a RREP packet reaches to the source, its ActiveNeighborCount field has already counted the exact number of this path's active neighbors. At this point, source can choose those RREPs from received ones that have the lowest ActiveNeighborCount and send information simultaneously through those paths. For this purpose, the source sets a clock after receiving the first RREP and waits for receiving the rest of RREP. After the timer expires, it chooses paths with less ActiveNeighborCount .

3.3. The suggested algorithm's pseudo-code

For a better understanding of the proposed algorithm, the source nodes function pseudo code in figure 2, destination node function pseudo code in figure 3, and pseudo code of the middle node function are presented in Figure 4.

1. If you have data to send and you don't have a valid path to that destination, broadcast the RREQ packet.
2. Wait for RREP to arrive.
3. In case of receiving the first RREP, wait for a while and then choose those paths among received paths that have the lowest number of neighbors. After wards, start sending data via this path.

Figure 2. Source node pseudo code in the suggested algorithm

1. Send the corresponding path PPEP for all the nodes that you have received the RREQ packets.

Figure 3. Destination node pseudo code in the proposed algorithm

1. If you received the RREQ packet and this packet is acceptable, do the following steps. Otherwise, dismiss the packet.
      a. Put this packet's specification into the RREQ_Seen table.
      b. Prepare the RREQ_QUERY packet and assign it a value.
      c. There is a question on this packet that asks: Have you seen such a request packet before?
      d. Send the RREQ_Query packet to your neighbors
      e. Wait a specific period of time for your neighbors to reply
      f. Increase the ActiveNeighborCount with regard to the number of accepted replies.
      g. Rebroadcast the RREQ packet
2. When you received the RREQ_Query packet, perform the following actions:
      a. With regard to the RREQ_Seen table, if you have not seen this RREQ before, dismiss the packet and don't consider it.
      b. According to the REQ_Seen table, if you have seen this RREQ before, inform the query node by sending a RREQ_Query_Reply packet then add one unite to the After_A_N_C field of the corresponding RREQ in its RREQ_Seen table.
3. If you have received the RREQ_Qeury_Reply packet, add one unite to this RREQ's AvtiveNeighborCount field.
4. When you receive the RREP packet, add the corresponding after_a_n_c to activeneighborcount field of RREP packet and send it.

Figure 4. middle node pseudo code in the proposed algorithm

For a better understanding of the algorithm, consider the hypothetical network in figure 5. In this network, the line between two nodes means that these two nodes are in each other's radio range.

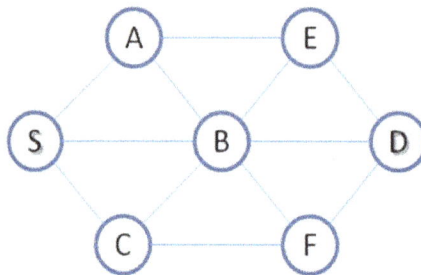

Figure 5. Network Topology of our example.

Suppose in this example, the node S as a source node wants to send data to node D as the destination but it does not know the path to the destination node. Therefore, node S sends the RREQ packet with the zero value of ActiveNeighborCount field to all. In the first stage, A, B and C receive the packet and insert its specifications in RREQ_Seen table along with the initial value of zero for After_A_N_C field. Then they add their address as the founder of a path in the RREQ packet. In addition, they begin the query procedure according to the proposed algorithm. For this purpose, they send the RREQ_Query packet for their neighbors and wait for their neighbors to respond by setting a clock. After making inquiries, node A and C only recognize the node B in their neighboring and add one unite to the ActiveNeighborCount field in RREQ Packet, but node B recognizes the neighboring of the node A and C. Also, it adds two units to ActiveNeighborCount field in RREQ Packet. Then all three nodes of A, B and C propagate the RREQ packet to complete the discovery process. This process is shown in figure 6.

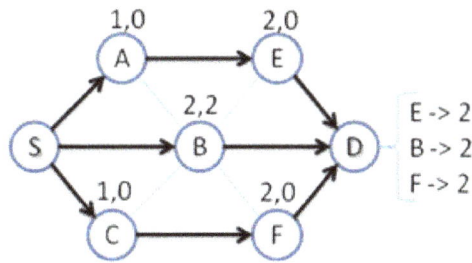

Figure 6. First Phase of our Algorithm.

Each node has two numbers written in its above, the number on the left shows the value of ActiveNeighborCount field and the number on the right shows the After_A_N_C field value exactly before re-broadcast RREQ in the corresponding RREQ_Seen table. In the next step, these RREQs reach other nodes. Because of the simplicity of the figures 6 and 7, we refused to show other versions of the RREQ which have been delivered to nodes by other paths. The node B delivers the first RREQ to D, and the destination node replies by sending a RREP packet to the source. Noticeably, this path's ActiveNeighborCount is equal to 2. However, RREQ reaches to the E and F through the paths in the corresponding figure. These two nodes begin the query processes separately, recognize the node B in their neighboring, and add one unite to the existing ActiveNeighborCount in RREQ. Factually, the nodes E and F send the RREQ_Query packets to node B to perform the query process. Also, because node B has propagated this RREQ before, it has a positive answer for both queries and it sends the RREQ_Query_Reply packet to each. Furthermore, node B finds two new neighbor (E and F)after this inquiry. Therefore, after sending the query reply, node B adds one unite to After_A_N_C field of current path in RREQ_Seen table per each query. This procedure is well shown in Figure 6, and we can see that the value of After_A_N_C field of node B is equal to 2. Figure 6 shows the network status at the moment that all RREQs have reached the destination. As it is indicated in the figure, all three RREQs have been received in destination with an equal amount of ActiveNeighborCount which is 2. At this stage, the destination receives the RREQ, creates its relevant RREP Packet, fills the corresponding RREQ with the same amount of the existing ActiveNeighborCount field in RREP, and then sends the RREP packets to source.

After receiving each of these RREPs, middle nodes must add the After_A_N_C value from RREQ_Seen ActiveNeighborCount table to field in RREP. This is shown in figure 7. In this figure, the corresponding sum action of each node is shown above the nodes. As it can be seen, nodes A, E, F and C in this scenario does not add any value to the ActiveNeighborCount field in RREP, but node B adds two units to ActiveNeighborCount in its own RREP. In the end, RREP packets are delivered to the source (node S).

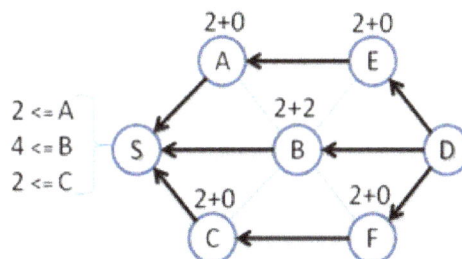

Figure 7. Second Phase of our Algorithm.

By receiving the first RREP and setting a timer, the source node waits for a certain period of time. After timer expiration, the source node sorts the received RREPs in an ascending order.

Then it selects the required number of RREPs from the beginning of queue and begins to send the information concurrently through the selected paths. (In fact, those paths are selected that have less ActiveNeighborCount.) For example, we suppose that the origin is determined to send data concurrently to the destination through two paths. In this case, as it is shown in figure 8, both the SAED and S-CFD paths can be selected.

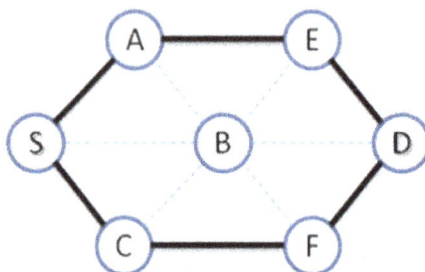

Figure 8. Selected Paths in our example

4. SIMULATION

The results of simulations and a comparison between the proposed algorithm and other existing algorithms are presented in this section. For this purpose, the following algorithms have been compared with each other in various scenarios:

- The proposed multi-path routing algorithm, which is presented as ZD-AOMDV in graphs and results.
- AOMDV [11]
- AODVM [12]
- IZM-DSR [14]

4.1. Simulation Environments

In this study, GLOMOSIM is used for the simulation [15]. For this purpose, we have compared the proposed algorithm with AOMDV algorithm in various scenarios. Conditions, simulation environment and simulation results are presented in this section. In these simulations, both algorithms use three paths for sending data simultaneously.

50 nodes with radio range of 250m in an environment with the dimension of 750x750m are used for simulation. In such status, nodes have random movement with using the Random Waypoint mobility model. In this model, each node randomly selects a point as a destination. After the node reaches to destination, it stays at the same point for the duration of Pause Time and again it repeats the same action. In all simulations, we consider one second for Pause Time. Nodes also use IEEE 802.11 MAC layer protocol; moreover, nodes use RADIO-ACCNOISE standard radio model for sending or receiving information. CBR model is used for traffic model. The time duration for each simulation has been considered 300 seconds and the recorded results is the outcome of an average 25 times for each simulation.

4.2. Simulation Metrics

Five important performance metrics were evaluated in our simulation: (i) End-to-End Delay Average –this includes all possible delays caused by buffering during route discovery phase, queuing at the interface queue, retransmission at the MAC layer, propagation and transfer delays. (ii) Packet Delivery Ratio, (iii) Routing Overhead Ratio – the number of routing control packets per each data packet. (iv) Energy Consumption. (v) Number of Dead Nodes.

4.3. Simulation results

A) Packet Delivery Ratio

In figures 9-12, the 4 algorithms have been compared with each other in terms of packet delivery rate according to the results of the simulation. In all algorithms, the packet delivery rate will decrease by increasing the maximum speed of nodes .Moreover, this decrease is because of the dynamic nature of topology and the increase of network connections termination rate.

Figure 9 shows packet delivery ratio versus max speed of nodes in random way point model, figure 10 shows packet delivery ratio versus pause time of nodes in random way point model, figure 11 shows packet delivery ratio versus number of data sources (number of connections) and figure 12 shows packet delivery ratio versus offered traffic.

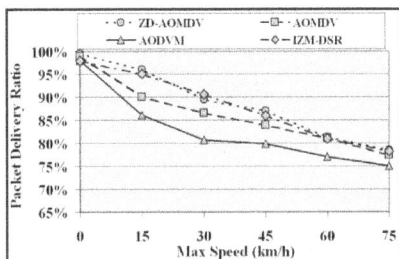

Figure 9. Packet delivery ratio vs. Max Speed.
(Pause time=15ms, No of Src=1, Traffic= 35Kbps)

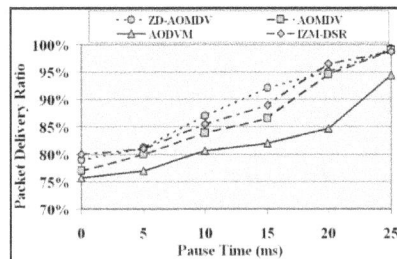

Figure 10. Packet delivery ratio vs. Pause time
(Max speed=40Km/h, No of Src=1, Traffic=35Kbps)

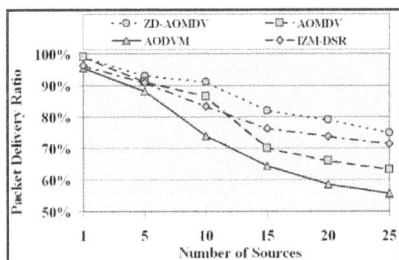

Figure 11. Packet delivery ratio vs. No. of Sources
(Max speed=40Km/h, Pause time=15ms,Trffic=35Kbps)

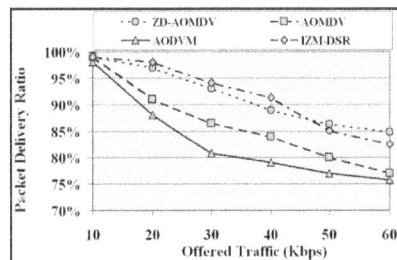

Figure 12. Packet delivery ratio vs. Offered Terrafic
(Max speed=40Km/s, Pause time=15ms, No of Src=1)

B) End-to-end Delay

In figures 13-16, the four algorithms have been compared in terms of the average of end-to-end delay. As it can be seen, with increasing the maximum speed of nodes the end-to-end delay average of packets increases correspondingly. With focusing on figure 13, the ZD-AOMDV algorithm reaches to less end-to-end delay than the other algorithms. As it has been described, the route discovery phase of the ZD-AOMDV algorithm is run with more delay. Instead, this algorithm compensate for this delay while it sends data. As a result, the end-to-end delay for algorithm ZD-AOMDV is less than other algorithms.

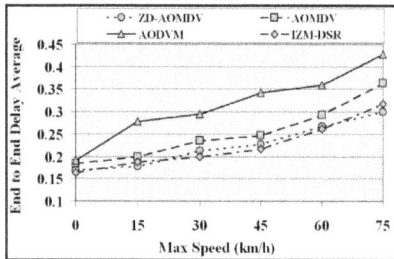

Figure 13. End-to-end delay vs. Max speed
(Pause time=15ms, No of Src=1, Traffic=35Kbps)

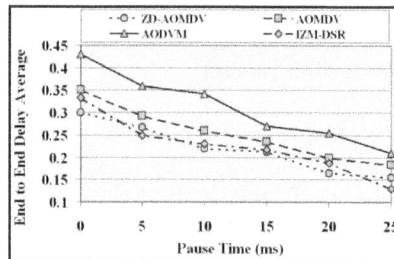

Figure 14. End-to-end delay vs. Pause time
(Max speed=40Km/h, No of Src=1, Traffic=35Kbps)

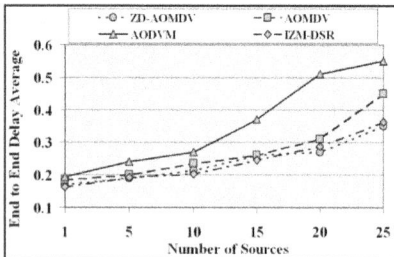

Figure 15. End-to-end delay vs. No. of Sources
(Max speed=40Km/h, Pause time=15ms, Traffic=35Kbps)

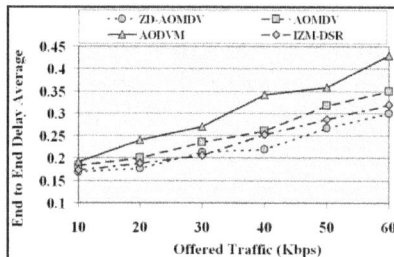

Figure 16. End-to-end delay vs. Offered Traffic
(Max speed=40Km/h, Pause time=15ms, No of Src=1)

C) Routing Overhead

Also in this section, the four algorithms have been compared in terms of routing overhead. The ZD-AOMDV and AOMDV algorithms send data through three paths simultaneously. The results of this simulation model have been displayed in figures 17-20. In this case the proposed algorithm has greater control overhead ratio than other algorithms.

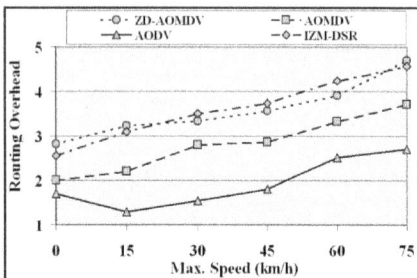

Figure 17. Routing overhead vs. Max speed
(Pause time=15ms, No of Src=1, Traffic=35Kbps)

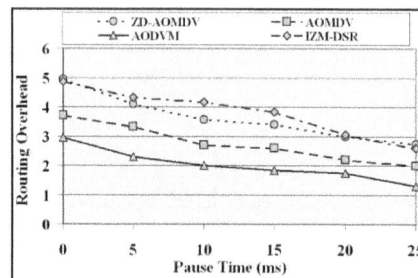

Figure 18. Routing overhead vs. Pause time
(Max speed=40Km/h, No of Src=1, Traffic=35Kbps)

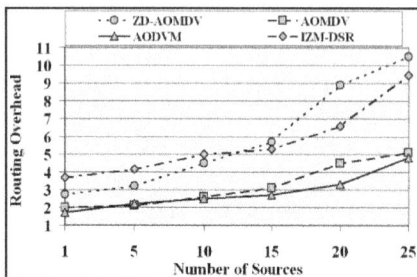

Figure 19. Routing overhead vs. No. of Sources
(Max speed=40Km/h, Pause time=15ms, Traffic=35Kbps)

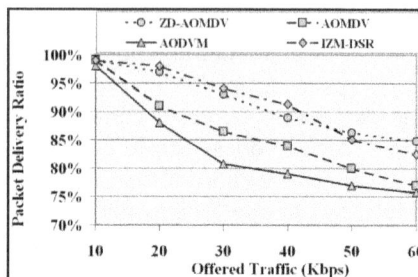

Figure 20. Routing overhead vs. Offered Traffic
(Max speed=40Km/h, Pause time=15ms, No of Src=1)

D) Number of Dead Nodes

In figure 23 the number of dead nodes in the network is depicted, and if we consider the network lifetime to be the period in which at least half of the nodes in the network are alive, then it can be realized from this figure that by using ZD-AOMDV instead of using AODVM protocol, network lifetime will nearly be doubled.

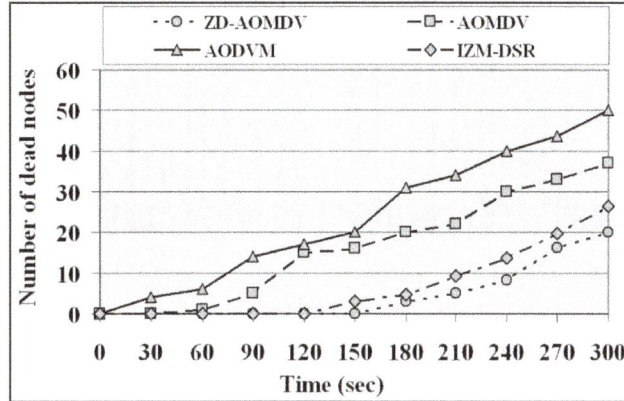

Figure 21. Number of Dead Nodes in Simulation Time.

E) Energy Consumption

In figures 22, 23 and 24 energy consumption are shown as a function of mobility speed and data rate. In ZD-AOMDV and IZM-DSR all discovered routes are used simultaneously thus many of nodes participate in forwarding data and their consumed energies are increased simultaneously, but in AODVM and AOMDV only one route is used for forwarding data thus only a few nodes are involved in transmitting data. In fact, ZD-AOMDV distributes energy consumption across many nodes thus the life time of network is increased but in AODVM and AOMDV the energy consumption is focused on a few nodes so the network life time is less than in ZD-AOMDV.

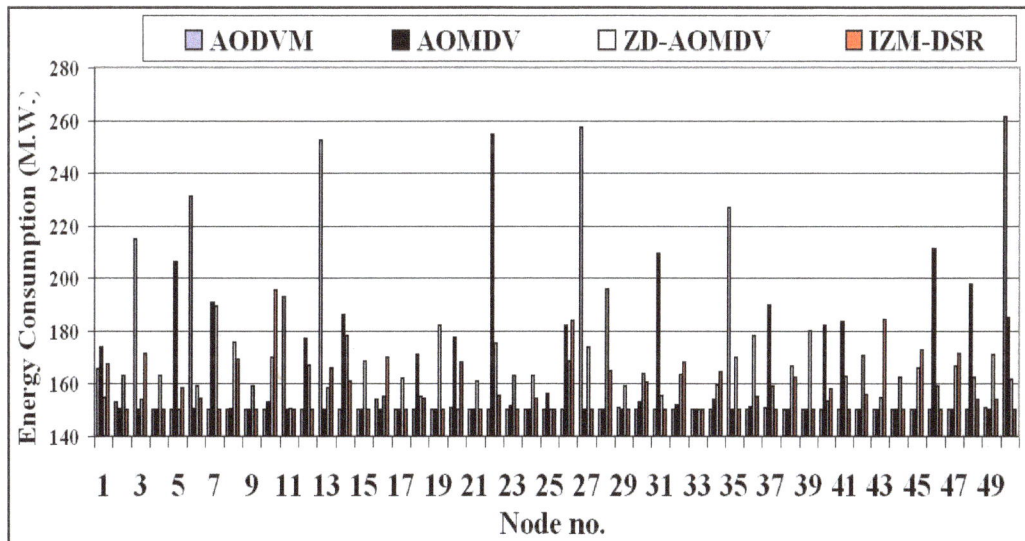

Figure 22. Energy Consumption
(Max speed=30Km/h, Pause time=15ms, No of Src=1, Terrafic=30Kbps)

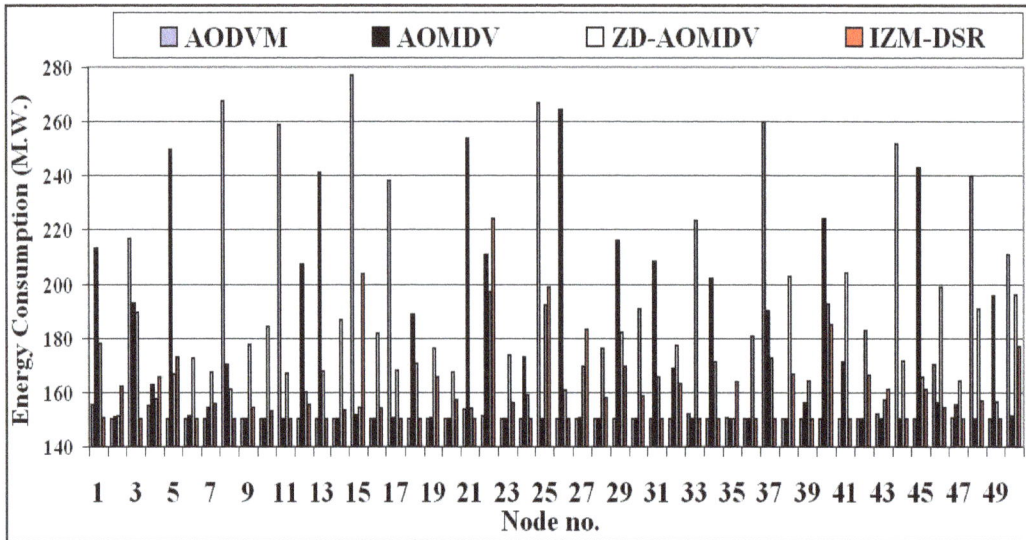

Figure 23. Energy Consumption
(Max speed=60Km/h, Pause time=30ms, No of Src=1, Terrafic=30Kbps)

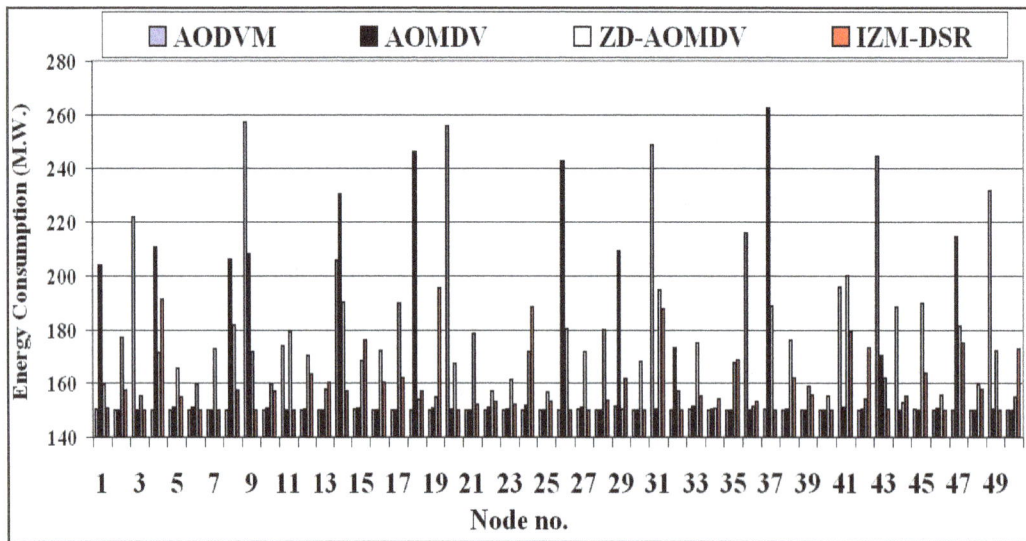

Figure 24. Energy Consumption
(Max speed=30Km/h, Pause time=15ms, No of Src=1, Terrafic=60Kbps)

With getting more thorough in figure 10, we can realize that the increase of the number of nodes has a very large effect on ZD-AOMDV overhead routing .This is because of the neighboring nodes increase and increase of the number of packets per query and query response in the path discovery process.

5. CONCLUSION

Some multipath algorithms in the ad hoc networks divide data at source and simultaneously send the different parts to destination via different paths to reduce end-to-end delay. In this way, using node disjoint paths seems like a good option. But sending traffic through node disjoint paths is not completely independent of each other and because of the mechanisms for shared channel access in wireless networks such as the CSMA/CA protocol, sending information through a path can affect another path. Such problems can be solved by implementing regional disjoint paths instead of node disjoint paths for sending information concurrently. In this paper, a new multipath routing algorithm is suggested based on AODV that uses all directional antenna to discover and use regional distinct paths. To achieve this goal, active neighbors of each path are counted. Also, selection is executed based on the number of active neighbors.

The proposed algorithm is compared to AOMDV, AODVM and IZM-DSR algorithms during various scenarios, and improvements are obtained in the field of energy consumption, end-to-end delay and packet delivery ratio. But instead, our proposed algorithm's routing overhead is higher than AOMDV and AODVM algorithms.

REFERENCES

[1]   S. Sesay, Z. Yang, and J. He, (2004), "A Survey on Mobile Ad Hoc Wireless Network," Information Technology Journal, vol. 2, pp: 168-175.

[2]   E. Royer, and C. Toh, (1999), "A Review of Current Routing Protocols for Ad-hoc Mobile Wireless Networks," IEEE Personal Communication Magazine, pp: 46-55.

[3]   S. Mueller, R. Tsang, and D. Ghosal, (2004), "Multipath Routing in Mobile Ad Hoc Networks: Issues and Challenges," Lecture Notes in Computer Science (LNCS 2965), pp: 209-234.

[4]   A. Colvin, (1983), "CSMA with Collision Avoidance," Computer Communication, Vol. 6, pp: 227-235.

[5]   N. Taheri Javan, and M. Dehghan, (2007), "Reducing End To End Delay in Multi-Path Routing Algorithms for Mobile Ad Hoc Networks," The 3rd International Conference on Mobile Ad-hoc and Sensor Networks, Beijing, China, Published in Lecture Notes in Computer Science, (LNCS 4864), Springer-Verlag, pp: 715-724.

[6]   S. Roy, D. Saha, S. Bandyopadhyay, T. Ueda, and S. Tanaka, (2003), "Improving End-to-End Delay through Load Balancing with Multipath Routing in Ad Hoc Wireless Networks using Directional Antenna," 5th International Workshop IWDC, LNCS, pp: 225-234.

[7]   N. Taheri Javan, R. KiaeeFar, B. Hakhamaneshi, and M. Dehghan, (2009), "ZD-AOMDV: A New Routing Algorithm for Mobile Ad hoc Networks," 8th International Conference on Computer and Information Science (ICIS), Shanghai, China, pp: 852-857.

[8]   L. Guang-cong, Z. Hua, and W. Dong-li, (2011), "Node-disjoint multi-path routing algorithm based on AODV in Ad hoc networks," Application Research of Computers, vol. 28, no. 2, pp: 692-695.

[9]   M. Nagaratna, C. Raghavendra, and V. K. Prasad, (2011), "Node Disjoint Split Multipath Multicast Ad-hoc On-demand Distance Vector Routing Protocol (NDSM-MAODV)," International Journal of Computer Applications, vol. 26, no. 10, pp: 1-12.

[10]   M. H. Shao, and Y. P. Lee, (2011), "An Adaptive Link-Disjoint Multipath Routing in Ad Hoc Networks," Advanced Materials Research, Vols. 171 - 172, pp: 628-63.

[11]   M. K. Marina, and S. R. Das, (2001), "On Demand Multipath Distance Vector Routing in Ad hoc Networks," IEEE International Conference on Network Protocols (ICNP), California, USA, pp: 14-23.

[12]   Z. Ye, S. Y. Krishnamurthy, and S. K. Tripathi, (2003), "A framework for reliable routing in mobile ad hoc networks," IEEE INFOCOM, San Francisco, pp: 270–280.

[13]   A. Dareshoorzadeh, N. Taheri Javan, M. Dehghan, and M. Khalili, (2008), "LBAODV: A New Load Balancing Multipath Routing Algorithm for Mobile Ad hoc Networks," 6th IEEE NCTT-MPC, Malaysia, pp: 344-349.

[14]   N. Taheri Javan, A. DareshoorZade, S. Soltanali, and Y. Ghanbari Birgani, (2009), "IZM-DSR: A New Zone-disjoint Multi-path Routing Algorithm for Mobile Ad hoc Networks," 3rd European Symposium on Computer Modeling and Simulation (EMS), Athens, Greece, pp: 511-516.

[15]   L. Bajaj, M. Takai, R. Ahuja, R. Bagrodia, and M. Gerla, (1999), "Glomosim: a Scalable Network Simulation Environment," Technical Report 990027, Computer Science Department, UCLA.

16

PERFORMANCE COMPARISONS OF ROUTING PROTOCOLS IN MOBILE AD HOC NETWORKS

P. Manickam[1], T. Guru Baskar [2], M.Girija[3], Dr.D.Manimegalai [4]

[1,2]Department of Applied Sciences, Sethu Institute of Technology, India
gm7576@gmail.com

[3]Department of Computer Science, The American College, India

[4]Department of Information Technology, National Engineering College, India

ABSTRACT

Mobile Ad hoc Network (MANET) is a collection of wireless mobile nodes that dynamically form a network temporarily without any support of central administration. Moreover, Every node in MANET moves arbitrarily making the multi-hop network topology to change randomly at unpredictable times. There are several familiar routing protocols like DSDV, AODV, DSR, etc... which have been proposed for providing communication among all the nodes in the network. This paper presents a performance comparison of proactive and reactive protocols DSDV, AODV and DSR based on metrics such as throughput, packet delivery ratio and average end-to-end delay by using the NS-2 simulator.

KEYWORDS

MANET, DSDV, AODV, DSR, Throughput, Packet Delivery Ratio, Average End-to-End delay

1. INTRODUCTION

A mobile ad hoc network is a collection of wireless mobile nodes that dynamically establishes the network in the absence of fixed infrastructure [1]. One of the distinctive features of MANET is, each node must be able to act as a router to find out the optimal path to forward a packet. As nodes may be mobile, entering and leaving the network, the topology of the network will change continuously. MANETs provide an emerging technology for civilian and military applications. Since the medium of the communication is wireless, only limited bandwidth is available. Another important constraint is energy due to the mobility of the nodes in nature.

One of the important research areas in MANET is establishing and maintaining the ad hoc network through the use of routing protocols. Though there are so many routing protocols available, this paper considers DSDV, AODV and DSR for performance comparisons due to its familiarity among all other protocols. These protocols are analyzed based on the important metrics such as throughput, packet delivery ratio and average end-to-end delay and is presented with the simulation results obtained by NS-2 simulator.

In particular, Section 2 presents the related works with a focus on the evaluation of the routing protocols. Section 3 briefly discusses the MANET routing protocols classification and the functionality of the three familiar routing protocols DSDV, AODV and DSR. The simulation results and performance comparison of the three above said routing protocols are discussed in Section 4. Finally, Section 5 concludes with the comparisons of the overall performance of the three protocols DSDV, AODV and DSR based on the throughput, packet delivery ratio and average end-to-end delay metrics.

2. RELATED WORK

A number of routing protocols have been proposed and implemented for MANETs in order to enhance the bandwidth utilization, higher throughputs, lesser overheads per packet, minimum consumption of energy and others. All these protocols have their own advantages and disadvantages under certain circumstances. The major requirements of a routing protocol was proposed by Zuraida Binti et al.[4] that includes minimum route acquisition delay, quick routing reconfiguration, loop-free routing, distributed routing approach, minimum control overhead and scalability.

MANET Routing Protocols possess two properties such as Qualitative properties (distributed operation, loop freedom, demand based routing & security) and Quantitative properties (end-to-end throughput, delay, route discovery time, memory byte requirement & network recovery time). Obviously, most of the routing protocols are qualitatively enabled. A lot of simulation studies were carried out in the paper [2] to review the quantitative properties of routing protocols.

A number of extensive simulation studies on various MANET routing protocols have been performed in terms of control overhead, memory overhead, time complexity, communication complexity, route discovery and route maintenance[16][4]. However, there is a severe lacking in implementation and operational experiences with existing MANET routing protocols. The various types of mobility models were identified and evaluated by Tracy Camp et al. [6] because the mobility of a node will also affect the overall performance of the routing protocols. A framework for the ad hoc routing protocols was proposed by Tao Lin et al. [3] using Relay Node Set which would be helpful for comparing the various routing protocols like AODV, OLSR & TBRPF [17].

The performance of the routing protocols OLSR, AODV and DSR was examined by considering the metrics of packet delivery ratio, control traffic overhead and route length by using NS-2 simulator [19][2][20][22]. The performance of the routing protocols OLSR, AODV, DSR and TORA was also evaluated with the metrics of packet delivery ratio, end-to-end delay, media access delay and throughput by also using OPNET simulator [21][23][18].

3. MOBILE AD HOC NETWORK ROUTING PROTOCOLS

3.1. Protocol Classifications

There are many ways to classify the MANET routing protocols (Figure 1), depending on how the protocols handle the packet to deliver from source to destination. But Routing protocols are broadly classified into three types such as Proactive, Reactive and Hybrid protocols [5].

3.1.1. Proactive Protocols

These types of protocols are called table driven protocols in which, the route to all the nodes is maintained in routing table. Packets are transferred over the predefined route specified in the routing table. In this scheme, the packet forwarding is done faster but the routing overhead is greater because all the routes have to be defined before transferring the packets. Proactive protocols have lower latency because all the routes are maintained at all the times.
Example protocols: DSDV, OLSR (Optimized Link State Routing)

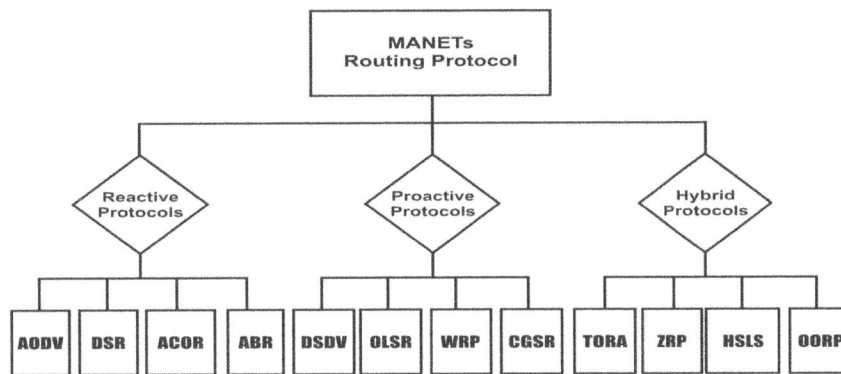

Figure 1. MANET Routing Protocols

3.1.2. Reactive Protocols

These types of protocols are also called as On Demand Routing Protocols where the routes are not predefined for routing. A Source node calls for the route discovery phase to determine a new route whenever a transmission is needed. This route discovery mechanism is based on flooding algorithm which employs on the technique that a node just broadcasts the packet to all of its neighbors and intermediate nodes just forward that packet to their neighbors. This is a repetitive technique until it reaches the destination. Reactive techniques have smaller routing overheads but higher latency.
Example Protocols: DSR, AODV

3.1.3. Hybrid Protocols

Hybrid protocols are the combinations of reactive and proactive protocols and takes advantages of these two protocols and as a result, routes are found quickly in the routing zone.
Example Protocol: ZRP (Zone Routing Protocol)

3.2. Overview of Routing Protocols

In this section, a brief overview of the routing operations performed by the familiar protocols DSDV, AODV and DSR are discussed.

3.2.1. Destination-Sequenced Distance-Vector (DSDV) protocol

The Table-driven DSDV protocol is a modified version of the Distributed Bellman-Ford (DBF) Algorithm that was used successfully in many dynamic packet switched networks [14]. The Bellman-Ford method provided a means of calculating the shortest paths from source to destination nodes, if the metrics (distance-vectors) to each link are known. DSDV uses this idea, but overcomes DBF's tendency to create routing loops by including a parameter called destination-sequence number.

In DSDV, each node is required to transmit a sequence number, which is periodically increased by two and transmitted along with any other routing update messages to all neighboring nodes. On reception of these update messages, the neighboring nodes use the following algorithm to decide whether to ignore the update or to make the necessary changes to its routing table:

Step 1: Receive the update message
Step 2: Update the routing table if any one of the following condition satisfies:
  i) $S_n > S_p$
  ii) $S_n = S_p$, Hop count is less
    Otherwise, ignore the update message.

Here, S_n and S_p are the Sequence numbers of new message and existing message respectively.

When a path becomes invalid, due to movement of nodes, the node that detected the broken link is required to inform the source, which simply erases the old path and searches for a new one for sending data. The advantages are latency for route discovery is low and loop-free path is guaranteed. The disadvantage is the huge volume of control messages.

3.2.2. Ad Hoc On-demand Distance Vector Routing (AODV) protocol

The Ad Hoc On-demand Distance Vector Routing (AODV) protocol is a reactive unicast routing protocol for mobile ad hoc networks [12]. As a reactive routing protocol, AODV only needs to maintain the routing information about the active paths. In AODV, the routing information is maintained in the routing tables at all the nodes. Every mobile node keeps a next-hop routing table, which contains the destinations to which it currently has a route. A routing table entry expires if it has not been used or reactivated for a pre-specified expiration time.

In AODV, when a source node wants to send packets to the destination but no route is available, it initiates a route discovery operation. In the route discovery operation, the source node broadcasts route request (RREQ) packets which includes Destination Sequence Number. When the destination or a node that has a route to the destination receives the RREQ, it checks the destination sequence numbers it currently knows and the one specified in the RREQ. To guarantee the freshness of the routing information, a route reply (RREP) packet is created and forwarded back to the source only if the destination sequence number is equal to or greater than the one specified in RREQ.

AODV uses only symmetric links and a RREP follows the reverse path of the respective RREQ. Upon receiving the RREP packet, each intermediate node along the route updates its next-hop table entries with respect to the destination node. The redundant RREP packets or RREP packets with lower destination sequence number will be dropped. The advantage of this protocol is low Connection setup delay and the disadvantage is more number of control overheads due to many route reply messages for single route request.

3.2.3. Dynamic Source Routing (DSR) Protocol

The Dynamic Source Routing (DSR) is a reactive unicast routing protocol that utilizes source routing algorithm [13]. In DSR, each node uses cache technology to maintain route information of all the nodes. There are two major phases in DSR such as:
* Route discovery
* Route maintenance

When a source node wants to send a packet, it first consults its route cache [7]. If the required route is available, the source node sends the packet along the path. Otherwise, the source node initiates a route discovery process by broadcasting route request packets. Receiving a route request packet, a node checks its route cache. If the node doesn't have routing information for the requested destination, it appends its own address to the route record field of the route request packet. Then, the request packet is forwarded to its neighbors.
If the route request packet reaches the destination or an intermediate node has routing information to the destination, a route reply packet is generated. When the route reply packet is

generated by the destination, it comprises addresses of nodes that have been traversed by the route request packet. Otherwise, the route reply packet comprises the addresses of nodes the route request packet has traversed concatenated with the route in the intermediate node's route cache.

Whenever the data link layer detects a link disconnection, a ROUTE_ERROR packet is sent backward to the source in order to maintain the route information. After receiving the ROUTE_ERROR packet, the source node initiates another route discovery operation. Additionally, all routes containing the broken link should be removed from the route caches of the immediate nodes when the ROUTE_ERROR packet is transmitted to the source. The advantage of this protocol is reduction of route discovery control overheads with the use of route cache and the disadvantage is the increasing size of packet header with route length due to source routing.

4. SIMULATION RESULTS AND PERFORMANCE COMPARISONS

4.1. Simulation Model

Network Simulator (Version 2.29), widely known as NS2, is simply an event driven simulation tool that has proved useful in studying the dynamic nature of communication networks. Simulation of wired as well as wireless network functions and protocols (e.g., routing algorithms, TCP, UDP) can be done using NS2.

A simulation study was carried out to evaluate the performance of MANET routing protocols such as DSDV, AODV and DSR based on the metrics throughput, packet delivery ratio and average end-to-end delay with the following parameters:

Parameter	Value
Radio model	TwoRay Ground
Protocols	DSDV,AODV,DSR
Traffic Source	Constant Bit Rate
Packet size	512 bytes
Max speed	10 m/s
Area	500 x 500
Number of nodes	50, 75, 100
Application	FTP
MAC	Mac/802_11
Simulation time (Sec)	20, 40, 60, 80 & 100

4.2. Throughput

It is the ratio of the total amount of data that reaches a receiver from a sender to the time it takes for the receiver to get the last packet. When comparing the routing throughput by each of the protocols, DSR has the high throughput. It measures of effectiveness of a routing protocol. The throughput values of DSDV, AODV and DSR Protocols for 50, 75 and 100 Nodes at Pause time 20s, 40s, 60s, 80s and 100s are noted in Table-1 and they are plotted on the different scales to best show the effects of varying throughput of the above routing protocols (Figures 2, 3 & 4).

Based on the simulation results, the throughput value of DSDV increases initially and reduces when the time increases. The throughput value of AODV slowly increases initially and maintains its value when the time increases. AODV performs well than DSDV since AODV is an on-demand protocol. The throughput value of DSR increases at lower pause time and grows as the time increases. Hence, DSR shows better performance with respect to throughput among these three protocols.

Pause Time (Sec.)	Protocol								
	DSDV			AODV			DSR		
	50 N	75 N	100 N	50 N	75 N	100 N	50 N	75 N	100 N
20	314933	304192	1738.67	598851	692565	691495	680597	680597	680597
40	326862	315232	90380.9	547095	581015	587314	575919	575991	579794
60	230359	207078	57521.5	474272	495703	499404	492096	490886	493155
80	260288	242423	127322	439949	455565	458831	451614	450615	452834
100	276990	260298	166929	419988	432664	435074	428177	426776	429315

Table 1. Comparison of Throughput

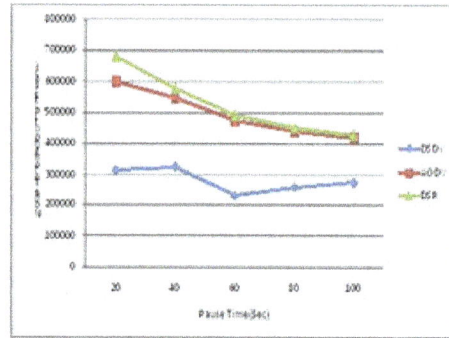

Figure 2. Comparison of Node Throughput for 50 Nodes

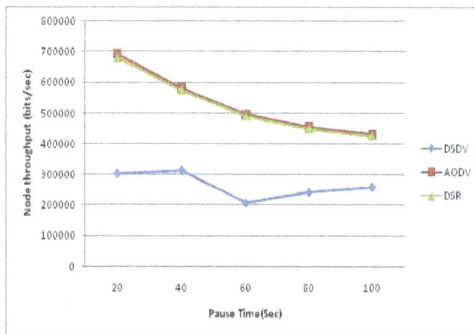

Figure 3. Comparison of Node Throughput for 75 Nodes

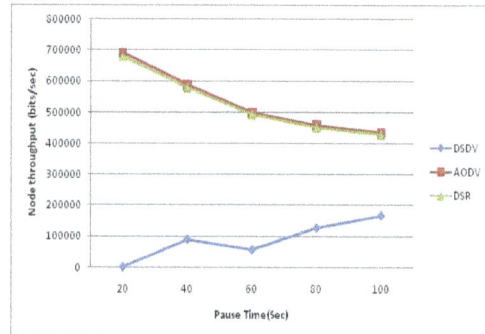

Figure 4. Comparison of Node Throughput for 100 Nodes

4.3. Packet delivery Ratio

Packet Delivery Ratio (PDR) is the ratio between the number of packets transmitted by a traffic source and the number of packets received by a traffic sink. It measures the loss rate as seen by transport protocols and as such, it characterizes both the correctness and efficiency of ad hoc routing protocols. A high packet delivery ratio is desired in any network.

The ratio of the Originated applications' data packets of each protocol which was able to deliver at varying time are shown in Figures 5,6 & 7 as per Table 2. As packet delivery ratio shows both the completeness and correctness of the routing protocol and also measure of efficiency the

Pause Time (Sec.)	Protocol								
	DSDV			AODV			DSR		
	50 N	75 N	100 N	50 N	75 N	100 N	50 N	75 N	100 N
20	97.6169	96.8661	80	99.0667	99.061	99.1886	99.1919	99.1909	99.1886
40	98.8569	98.5653	96.6102	99.1201	99.1093	99.1795	99.2424	99.2213	99.2031
60	98.4053	98.1191	96.4844	99.3528	99.3466	99.3864	99.4335	99.4166	99.404
80	98.8518	97.9306	97.2525	99.488	99.4843	99.5086	99.5467	99.5335	99.5233
100	98.4413	98.0971	97.4224	99.5764	99.5739	99.5907	99.6223	99.6113	99.6028

Table 2. Packet Delivery Ratio

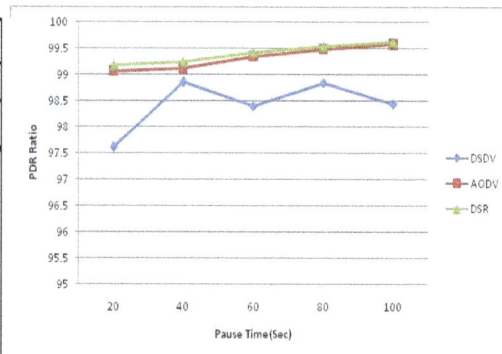

Figure 5. Comparison of PDR for 50 Nodes

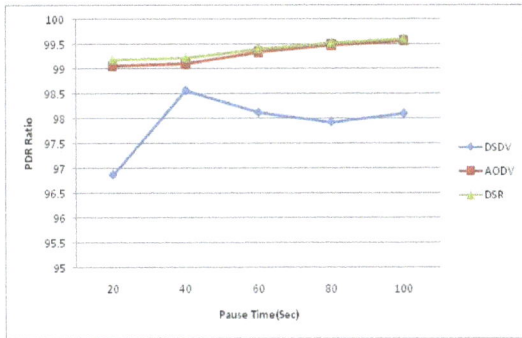

Figure 6. Comparison of PDR for 75 Nodes   Figure 7. Comparison of PDR for 100 Nodes

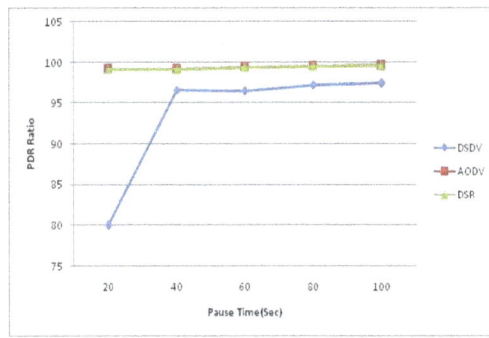

PDR value of AODV is higher than all other protocols. The PDR values of DSR and AODV are higher than that of DSDV. The PDR value of DSDV is worse in lower pause time and gradually grows in higher pause time. From the above study, in view of packet delivery ratio, reliability of AODV and DSR protocols is greater than DSDV protocol.

4.4. Average End-to-End delay

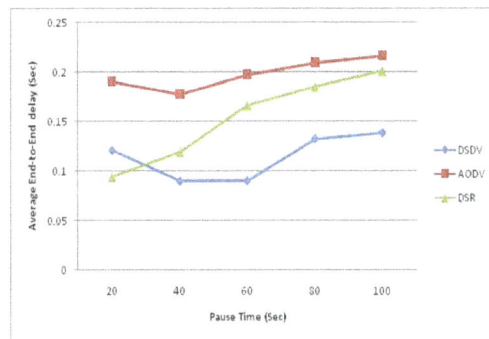

The packet End-to-End delay is the average time that a packet takes to traverse the network. This is the time from the generation of the packet in the sender up to its reception at the destination's application layer and it is measured in seconds. It therefore includes all the delays in the network such as buffer queues, transmission time and delays induced by routing activities and MAC control exchanges.

Various applications require different levels of packet delay. Delay sensitive applications such as voice require a low average delay in the network whereas other applications such as FTP may be tolerant to delays up to a certain level. MANETs are characterized by node mobility, packet retransmissions due to weak signal strengths between nodes, and connection tearing and making. These cause the delay in the network to increase. The End-to-End delay is therefore a measure of how well a routing protocol adapts to the various constraints in the network and represents the reliability of the routing protocol.

The Figures 8,9 &10 depict the average End-to-End delay for the DSDV, AODV and DSR protocols for the number of nodes 50, 75 & 100 respectively as per Table 3. It is clear that DSDV has the shortest End-to-End delay than AODV and DSR, because DSDV is a proactive protocol i.e. all routing informations are   already stored in table. Hence, it consumes lesser time

Pause Time (Sec.)	Protocol								
	DSDV			AODV			DSR		
	50 N	75 N	100 N	50 N	75 N	100 N	50 N	75 N	100 N
20	0.12090	0.12271	0.32929	0.19027	0.15404	0.17863	0.09408	0.16907	0.08187
40	0.08295	0.11878	0.12486	0.17764	0.15607	0.17468	0.11929	0.16157	0.10740
60	0.09035	0.11878	0.16703	0.19782	0.17982	0.19390	0.16596	0.18714	0.13623
80	0.13211	0.14868	0.24473	0.20844	0.19398	0.20469	0.18486	0.20473	0.13837
100	0.13818	0.15047	0.23451	0.21646	0.20857	0.21308	0.20101	0.22017	0.14495

Table 3. Average End-to-End delay   Figure 8. Comparison of Average End-to-
                                                              End delay for 50 Nodes

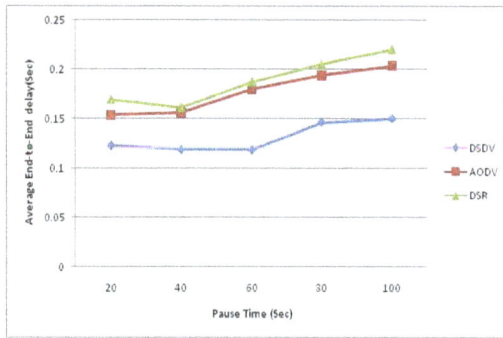

Figure 9. Comparison of Average End-to-End delay for 75 Nodes

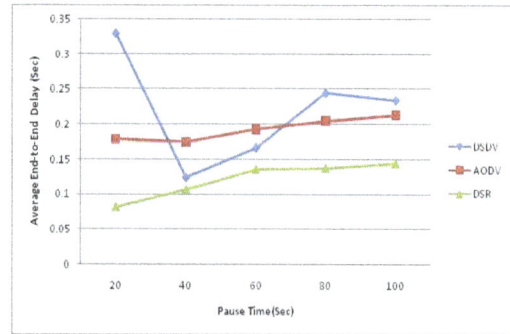

Figure 10. Comparison of Average End-to-End delay for 100 Nodes

than others. On average case, DSR shows better performance than AODV but worse than DSDV. As AODV needs more time in route discovery, it produces more End-to-End delay. From the above study on End-to-End delay, DSDV has high reliability than AODV and DSR.

5. CONCLUSION

In this paper, the performance of the three MANET Routing protocols such as DSDV, AODV and DSR was analyzed using NS-2 Simulator. We have done comprehensive simulation results of Average End-to-End delay, throughput, and packet delivery ratio over the routing protocols DSDV, DSR and AODV by varying network size, simulation time. DSDV is a proactive routing protocol and suitable for limited number of nodes with low mobility due to the storage of routing information in the routing table at each node. Comparing DSR with DSDV and AODV protocol, byte overhead in each packet will increase whenever network topology changes since DSR protocol uses source routing and route cache. Hence, DSR is preferable for moderate traffic with moderate mobility. As AODV routing protocol needs to find route by on demand, End-to-End delay will be higher than other protocols. DSDV produces low end-to-end delay compared to other protocols. When the network load is low, AODV performs better in case of packet delivery ratio but it performs badly in terms of average End-to-End delay and throughput. Overall, DSR outperforms AODV because it has less routing overhead when nodes have high mobility considering the above said three metrics.

REFERENCES

[1]   C.Sivaram murthy, B.S.Manoj, *Adhoc wireless networks:Architectures, and protocols*, Pearson Education, 2004.

[2]   Mohammed Bouhorma, H.Bentaouit and A.Boudhir, "Performance comparison of Ad hoc Routing protocols AODV and DSR" ,IEEE 2009.

[3]   Tao Lin, Scott F.Midkiff and Jahng S.Park ,"A framework for Wireless Ad hoc Routing Protcols", IEEE 2003.

[4]   Zuraida Binti Abdullah Hani and Mohd. Dani Bin Baba, "Designing Routing protocols for Mobile Ad hoc networks",IEEE 2003.

[5]   Mehran Abolhasan, Tadeusz Wysocki and Eryk Dutkiewicz ," A review of routing protocols for mobile ad hoc networks", Elsevier 2003.

[6]   Tracy Camp, Jeff Boleng and Vanessa Davies, " A Survey of Mobility Models for Ad Hoc Networks Research", Wireless communiaion & Mobile Computing (WCMC) 2002.

[7]   Yih-Chun Hu and David B.Johnson ",Caching Strategies in On-Demand Routing Protocols for Wireless Ad hoc Networks",ACM 2000.

[8]   H.M. El-Sayed,O.Bazon and U.Qureshi and M.Jaseemuddin "Performance Evaluation of TCP in Mobile Ad –Hoc Networks".

[9]   Manoj B.S., *Ad hoc Wireless Networks: Architectures and Protocols*, Prentice Hall of India, 2004.

[10]   Mansoor Mohsin and Ravi Prakash, "IP Address Assignment in a Mobile Ad Hoc Network", University of Taxas at Dallas, 2002.

[11]   Sanket Nesargi and Ravi Prakash, "MANETconf: Configuration of Hosts in a Mobile Ad Hoc Network", INFOCOM 2002."

[12]   C.Perkins, "Ad hoc on-demand distance vector (AODV) routing" ,RFC 3561,July 2003

[13]   D.Johnson, "The Dynamic Source Routing Protocol (DSR)", RFC4728, Feb 2007.

[14]   C.Perkins ,Praving Bhagwat, "Highly dynamic destination sequenced distance vector routing (DSDV) for Mobile computers".

[15]   Dhiraj Nitnaware, Ajay Verma, "Energy constraint Node cache based routing protocol for Adhoc Network", IJWMN, Feb. 2010.

[16]   Mehran Abolhasan, Tadeusz Wysoci, Eryk Dutkiewicz, "A review of routing protocols for mobile ad hoc networks", ELSEVIER , 2003.

[17]   Changling Liu, Jorg Kaiser, "A survey of Mobile Ad Hoc Network Routing Protocols", University of Magdeburg, 2005.

[18]   Md. Golam Kaosar, Hafiz M. Asif, Tarek R. Sheltami, Ashraf s. Hasan Mahmoud, "Simulation-Based Comparative Study of On-Demand Routing Protocols for MANET".

[19]   Thomas Heide Clausen, Phillippe Jacquet and Laurent Viennot, "Comparative Study of Routing Protocols for Mobile Ad-hoc Networks".

[20]   Saiful Azadm, Arafatur Rahman and Farhat Anwar, "A Performance comparison of Proactive and Reactive Routing protocols of Mobile Ad hoc Networks(MANET))", Journal of Engineering and Applied Sciences, 2007.

[21]   Nadia Qasim, Fatin Said and Hamid Aghvami, "Mobile Ad hoc Networks simulations using Routing protocols for Performance comparisons", Proceedings of the world congress on Enginccring, WCE, VOL I, 2008.

[22]   Wang Lin-zhu, FANG Ya-qin and SHAN Min, "Performance comparison of Two Routing Protocols for Ad Hoc Networks", WASE International conference on Information Engineering, 2009.

[23]   C.Mbarushimana and A.Shahrabi, "Comparative study of Reactive and Proactive Routing protocols performance in Mobile Ad Hoc Networks", AINAW-IEEE, 2007.

Authors

P. Manickam is working as an Assistant Professor in Sethu Institute of Technology, Tamil nadu, India. His current research focuses on Routing in Mobile Ad hoc Networks.

T. GuruBaskar is working as an Assistant Professor in Sethu Institute of Technology, Tamil nadu, India. His current research focuses on Security in Mobile Ad hoc Networks.

M. Girija is a Lecturer in The American College, Tamil nadu, India. Her area of specialization is Multicasting in Mobile Ad hoc Networks.

Dr.D.Manimegalai is Professor and Head, Department of Information Technology, National Engineering college, Tamil nadu, India. She has published more than fifteen research papers in national and International Journal and Conferences. Her area of specializations includes Image Processing, Web mining and Mobile Ad hoc Networks.

17

Optimization of Energy Consumption for OLSR Routing Protocol in MANET

Kirti Aniruddha Adoni[1] and Radhika D. Joshi[2]

[1]Modern College of Engineering, Shivaji Nagar, University of Pune, , India
akirti2008@gmail.com, k_adoni@rediffmail.com
[2]College of Engineering, University of Pune, India
rdj.extc@coep.ac.in

ABSTRACT

Mobile Adhoc Network (MANET) eliminates the complexity associated with an infrastructure networks. Wireless devices are allowed to communicate on the fly for applications. It does not rely on base station to coordinate the flow of the nodes in the network. This paper introduces an algorithm of multipath OLSR (Optimized Link State Routing) for energy optimization of the nodes in the network. It is concluded that this solution improves the number of nodes alive by about 10 to 25% by always choosing energy optimized paths in the network with some increase in normalized routing overheads.

KEYWORDS

OLSR, multipath, Energy Optimization, Nodes Alive, Overheads

1. INTRODUCTION

A Mobile Adhoc NETwork (MANET) is a multi-hop, distributed and self configuration network[1].The communication between two distant nodes is through the number of intermediate nodes which relays the information from one point to another. As nodes can move randomly within the network, routing packets between any pair of nodes become a challenging task. A route that is believed to be optimal for energy utilization at certain time might not be optimal at all, few moments later.[4]

Traditional proactive routing protocols[3,5] maintain routes to all nodes. Even if traffic is unchanged, repeated topology interaction happens among nodes. Also, they require periodic control message to maintain routes to every node in the network. Optimized Link State Routing (OLSR) is such a proactive routing protocol. Requirement of bandwidth and energy will increase for higher mobility . The behaviour of routing protocol depends on the network size and node mobility.

OLSR is an optimization of pure link state routing protocol which inherits the stability of a link state algorithm and takes over the advantage of proactive routing nature to provide routes immediately when needed. Here, to achieve energy optimization of all nodes in the network; first OLSR has been modified to multipath OLSR.

Among these multiple paths between the two distant nodes at given time, path containing all intermediate nodes with higher energies are considered.

The remaining of the paper is organized as follows. Section II discusses overview of OLSR routing protocol. Section III describes algorithm used for multipath and energy optimization in OLSRM by modification made in OLSR. Section IV describes simulation parameters to analyse performance differences. Section V discusses results of the OLSR and OLSRM for parameters like nodes alive and end to end delay, considering the effect of node velocity, node density and pause time. Finally, conclusions are in Section VI.

2. OVERVIEW OF OLSR ROUTING PROTOCOL

OLSR, proactive routing protocol exchanges routing information with other nodes in the network. The key concept used in OLSR is of MPRs (Multi Point Relays)[6]. It is optimized to reduce the number of control packets required for the data transmission using MPRs. To forward data traffic, a node selects its one hop symmetric neighbours, termed as MPR set that covers all nodes that are two hops away. In OLSR, only nodes, selected as MPRs are responsible for forwarding control traffic. The selected MPRs forward broadcast messages during the flooding process., contrarily to the classical link state algorithm, where all nodes forward broadcast messages. So mobile nodes can reduce battery consumption in OLSR compared with other link state algorithms.

2.1. Control Message

There are three types of control messages: HELLO messages, Topology Control (TC) messages, Multiple Interface Declaration MID messages. To achieve energy optimized multipath OLSR, HELLO message and TC message format has been modified.

- The link status and one hop neighbours' information data is given by HELLO message.

The format of a HELLO message is as follows [5]:

```
+-+-+-+-+-+-+-+-+-+-+-+-+-+-+-+-+-+-+-+-+-+-+-+-+
|      Reserved           |   Htime  |Willingness |
+-+-+-+-+-+-+-+-+-+-+-+-+-+-+-+-+-+-+-+-+-+-+-+-+
| Link Code  | Reserved  |    Link Message Size  |
+-+-+-+-+-+-+-+-+-+-+-+-+-+-+-+-+-+-+-+-+-+-+-+-+
|           Neighbor Interface Address          |
+-+-+-+-+-+-+-+-+-+-+-+-+-+-+-+-+-+-+-+-+-+-+-+-+
|           Neighbor Interface Address          |
+-+-+-+-+-+-+-+-+-+-+-+-+-+-+-+-+-+-+-+-+-+-+-+-+
:                  .  :                        :
+-+-+-+-+-+-+-+-+-+-+-+-+-+-+-+-+-+-+-+-+-+-+-+-+-|(etc):
```

Reserved: This field is always set to "0000000000000" by default. [This field is used to pass residual energy which will be useful for on hop neighbour while using OLSRM protocol.]

HTime: This field specifies the HELLO emission interval used by the node on this particular interface, i.e. the time before the transmission of the next HELLO(this information may be used in advanced link sensing).

Willingness: This field specifies the willingness of a node to carry and forward traffic for other nodes. A node with willingness WILL_NEVER, MUST never be selected as MPR by any node. A node with willingness WILL_ALWAYS MUST always be selected as MPR. By default, a node advertise a willingness of WILL_DEFAULT.

- Topology information is received by a node by periodical TC message using Multipoint Relaying (MPR) mechanism.

The format of a TC message is as follows[5]:

```
+-+-+-+-+-+-+-+-+-+-+-+-+-+-+-+-+-+-+-+-+-+-+-+-+-+
|           ANSN          |         Reserved       |
+-+-+-+-+-+-+-+-+-+-+-+-+-+-+-+-+-+-+-+-+-+-+-+-+-+
|            Advertised Neighbor Main Address      |
+-+-+-+-+-+-+-+-+-+-+-+-+-+-+-+-+-+-+-+-+-+-+-+-+-+
|            Advertised Neighbor Main Address      |
+-+-+-+-+-+-+-+-+-+-+-+-+-+-+-+-+-+-+-+-+-+-+-+-+-+
|                    ...                           |
+-+-+-+-+-+-+-+-+-+-+-+-+-+-+-+-+-+-+-+-+-+-+-+-+-+
```

Reserved

This field is reserved, and always be set to "0000000000000000" by default.

- MID message is sent on network to announce that if node is running multiple interfaces.

2.2. Routing table and Topology table

As a proactive routing, the routing table has routes for all available nodes in the networks. It has Destination Address, Next Hop Address, Local interface address and number of hops. It is as presented as follows:

Dest	next	iface	dist
0	0	37	1
2	6	37	2
14	20	37	3

From the above table, distance between 37 and 0 is 1 hop, the path is 37-0, distance between 37 and 2 is two hop, the path is 37-6-2, distance between 37 and 14 is 3 hop, the path is 37-20-?-14. If the number of hops are more than two, then intermediated nodes on the path has to find out the next (?), which is not displayed in routing table.

The topology table gives the information about entire network. It informs about one hop . There is no information about Residual energy of the node in topology table format of OLSR.

Its original format is as follows:

Dest	Last	Seq
0	13	2
36	13	2
1	13	2
0	38	6

2.3. Routing Discovery

To work in distributed manner, OLSR does not depend on any central entity [5]. Each node chooses its as multipoint relays (MPR) which are responsible to forward control traffics by flooding. The nodes maintain the network topology information where MPRs provide a shortest path to a destination with declaration and exchange of the link information periodically for their MPR's selectors. The HELLO messages are broadcast periodically for neighbour's detection and MPR selection process. It contains how often node send HELLO messages. It also includes node's MPR willingness and information about neighbour node. The information of node's is in the form of its link type, interface address and neighbour type .

The neighbour type can be one of: symmetric, MPR or not a neighbour. Link type indicates whether link is symmetric, asymmetric or lost link. A node is chosen as MPR if link to the neighbour is symmetric.

A node builds a one hop routing table with the reception of HELLO message information. It discards duplicate packet with same sequence number. The node updates when there is change in neighbour r node or route to a destination has expired.

OLSR does not require sequenced delivery of messages as each control message contains a sequence number which is incremented for each message.

2.4. Source Routing

Multiple paths calculated between a pair of source destination are independent, and they have no common nodes. However, because of the characteristic of next hop routing in OLSR, node can forward data based on its own routing table, and it cannot get the correct next node, source will forward, so cross among multiple paths happens. To avoid the problem for the next-hop routing in standard OLSR protocol, we use the source path in our multipath_OLSR algorithm. When a node calculates a path, the information of the path is recorded in its routing table (R_dest, R_next, Rdist, Rbuffer, nexthopO, and nexthopI... nexthopl4}. So, when source send data along the path, it add the source path (nexthopO, nexthopI... nexthopl4) to the IP header in the data. Now the intermediate nodes only need to get the path information from IP header of data to forward the data, need not to query its routing table as in standard OLSR protocol. So, the mechanism of source path added to multipath OLSR can avoid the problem of next hop node.

2.5. Energy-Efficient Route selection metric

There are different Route selection metric based on transmission power, link distance or residual energy of the node.
    A brief description of the relevant energy aware metric proposed are given below.

1. MTPR (Minimum Total Transmission Power Routing)[8]

The MTPR mechanism uses a simple energy metric. It repr esents the total energy consumed to forward the information along the route. MTPR uses shortest path routing. It reduces the overall transmission power consumed per packet. It does not take into account available residual energy of the node.

2. MBCR (Minimum Battery Cost Routing)[8]

The MBCR selects the route that minimizes the battery cost function. Battery cost function for a node is the reciprocal of available Residual energy of that node.

$$f(n_i) = \frac{1}{c(n_i)} \qquad (1)$$

Where $c(n_i)$ denotes the residual energy of node n_i.
Therefore, the battery cost for a route l, length D, is given by:

$$P_l = \sum_{i=0}^{D-1} f(n_i) \qquad (2)$$

The selected route P_k is the one that satisfies the following property,

$$P_k = min\{P_l : l \in A\} \qquad (3)$$

Where A is the set of all the possible routes.

The main disadvantage of the MBCR is that selection is based only on the battery cost. In this one node may be overused.

3. MMBCR (Min-Max Battery Cost Routing) [8]

The MMBCR selects the route with the maximum values of the minimum battery cost of the nodes. Therefore, the equation for battery cost is modified to,

$$P_l = \max_{i \in route\ i} f(n_i) \qquad (4)$$

The selected route P_k is the one that satisfies the following property:

$$P_k = min\{P_l : l \in A\} \qquad (5)$$

4. CMMBCR (Conditional Min-Max Battery Cost Routing) [8]

This mechanism considers both the total transmission power consumption of routes and the residual energy of nodes. When all nodes in some possible routes have sufficient remaining battery cost, i.e. above a threshold [criteria for setting the threshold based on application are subjective], MTPR is applied, to find out optimal path.

But, if all routes have nodes with low battery, i.e. below defined threshold, then MMBCR technique is applied. The performance of CMMBCR totally depends on selected threshold value.

5. MDR (Minimum Drain Rate)[8]

Only the Residual energy cannot be used to establish the best route between source and destination nodes. If a node has higher residual energy, too much traffic load will be injected through it, results in unfair sharp reduction of battery power. To avoid this problem MDR is used.

In this metric, cost function is considering both Residual energy of node and Drain rate of that node. Maximum Lifetime for a given path is determined by minimum value of cost along that path. Finally, MDR selects the optimal path having the highest maximum lifetime value.

6. LCMMER (Low Cost Min-Max Energy Routing) [9]

The difference between MMBCR and LCMMER is that MMBCR avoids the path with lowest energy nodes, does not consider the cost of the path and may select excessively long paths, whereas LCMMER also tries to avoid least energy nodes.

3. MODIFIED OLSR

OLSR applies shortest hop routing method for the transmission of data. It leads the congestion on specific path, or rise in energy expenditure of particular intermediate nodes.

If multiple paths are available, then congestion can be avoided, and energy expenditure of all nodes would be uniform. To achieve this, following changes are carried out.

Following are the changes made in OLSR protocol:

A. Changes in control messages

The 'reserved' field available in HELLO and TC message format is used to pass residual energy. This residual energy is further used to find out appropriate path.

Modified HELLO message format:

```
+-+-+-+-+-+-+-+-+-+-+-+-+-+-+-+-+-+-+-+-+-+-+-+-+-+
|   Residual Energy    |   Htime    |Willingness |
+-+-+-+-+-+-+-+-+-+-+-+-+-+-+-+-+-+-+-+-+-+-+-+-+-+
| Link Code  | Reserved |    Link Message Size    |
+-+-+-+-+-+-+-+-+-+-+-+-+-+-+-+-+-+-+-+-+-+-+-+-+-+
|            Neighbor Interface Address           |
+-+-+-+-+-+-+-+-+-+-+-+-+-+-+-+-+-+-+-+-+-+-+-+-+-+
|            Neighbor Interface Address           |
+-+-+-+-+-+-+-+-+-+-+-+-+-+-+-+-+-+-+-+-+-+-+-+-+-+
:                .  :                           :
+-+-+-+-+-+-+-+-+-+-+-+-+-+-+-+-+-+-+-+-+-+-+-+-|(etc)
```

Modified TC message format:

```
+-+-+-+-+-+-+-+-+-+-+-+-+-+-+-+-+-+-+-+-+-+-+-+-+
|        ANSN        |    Residual Energy      |
+-+-+-+-+-+-+-+-+-+-+-+-+-+-+-+-+-+-+-+-+-+-+-+-+
|         Advertised Neighbor Main Address      |
+-+-+-+-+-+-+-+-+-+-+-+-+-+-+-+-+-+-+-+-+-+-+-+-+
|         Advertised Neighbor Main Address      |
+-+-+-+-+-+-+-+-+-+-+-+-+-+-+-+-+-+-+-+-+-+-+-+-+
|                 ...                           |
+-+-+-+-+-+-+-+-+-+-+-+-+-+-+-+-+-+-+-+-+-+-+-+-+(etc)
```

B. Changes in Routing table and Topology table

As discussed in section II-B, in OLSR, user is not aware of intermediate nodes present on the path and also its residual energies.

The modified Routing Table of Multipath OLSR is as follows- from the modified routing table, information of residual energies of intermediate nodes are obtained.

```
Dest   next  iface   dist
14     20    37      3
20
11
14
Residual energy of intermediate node1
Residual energy of intermediate node2
.......
```

So from the modified Routing Table, for the given source-destination pair, multiple paths are available.

Now to select one of the available path, energy aware metric is applied.

The energy expenditure (in Joules) needed to transmit a packet p is given by,[7]

$$E(p) = i * v * t_p \qquad (6)$$

Where i is the current value,

      v is the voltage,

      t_p the time taken to transmit the packet p.

For our simulation, the voltage is chosen as 5 V.

Algorithm for modified OLSR:

- Maintain all one hop ing nodes for each node using modified HELLO message, with the residual energy of the nodes.
- Based on its one hop table, insert the appropriate entries to its routing table.
- Match the entries with topology set and add to the routing table.
- For each node, see recursively its last address until reached to the destination node, record the complete path information in the routing table using modified TC message (with the residual energy of the nodes).
- Discard the loop entries.
- Get all the paths for given source-destination pair, with the residual energy of each node to the entire network.
- Select all paths, for given source-destination pair
- Find out minimum energy of node, E(min), on each selected paths.
- Find out maximum energy of node, E(max), out of that E(min) values.
- Use this selected path.

4. SIMULATION PARAMETERS

We use network simulator ns2 [2] to analyse OLSR and OLSRM routing protocols and measure Number of node alive and Average end to end delay with varying Nodes' velocity and node density.

We use a movement pattern of the random waypoint mobility model, obtained from a tool called "setdest", developed by Carneige Mellon University. The performance of ad-hoc routing protocols greatly depends on the mobility model it runs over[10]. For simulation, Two ray ground propagation model is used. The nodes are 40 in the area of 1000 X 1000 square meter.

Traffic type used is CBR (Constant Bit Rate), Packet send rate is 20 packets/sec and Packet size is 512 Bytes.

5. SIMULATION RESULTS

Quality of Services (QoS) Parameters:

We evaluate essential Quality of Service parameters to analyse the performance differences of OLSR and OLSRM. Each node in the network has some constant Initial energy. The QoS parameter, alive nodes are chosen to show that more number of nodes alive for longer time in the network. More number of alive nodes implies the optimization of energy. The parameter Delay is chosen to study the effect of, addition of multipath technique and energy aware metric to the original OLSR

Number of Nodes Alive: This is one of the important metric to evaluate the energy efficiency of the routing protocol. It tells about Network Lifetime,
- The time to the first node failure due to battery outage
- The time to the unavailability of an application functionality[11]

First point gives the failure of first node, whereas second gives the time when only one node is alive (for communication at least two nodes must be alive). Both can be extracted from the trace file and tells about time at which first node died and the information about how alive nodes changes with the simulation time.

Average End to End Delay: This is the time difference between sending of data packets and time at which the same data packet is received .

Normalized Routing Overhead: It is the ratio of total number of routing packets to the total number of delivered data packets.

A. Effect of Node Mobility and Node Density on Number of Nodes Alive:

In case of OLSR, the shortest hop path is always chosen; whereas in OLSRM the path for the data delivery is considered with the available energy of nodes (at that instant) on the path, even if the path is long (in terms hop). Therefore OLSRM has more number of nodes alive compared to OLSR. As the node mobility increases, the number of alive nodes in OLSRM increases implies that modified protocol is suitable for dynamic network.

By varying number of nodes, it has been observed that OLSRM has more number of nodes, for high node density. It is obvious, as multiple paths will be more for large number of nodes. So it can be seen from the results, OLSRM is best suitable for dynamic and dense network.

Nodes Alive Vs Maximum Number of Nodes [At sim. Time 988s, Total Sim. Time 1000s]

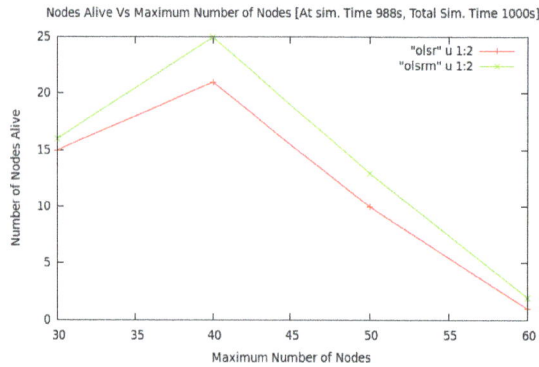

Fig. 1: Effect of node density on nodes alive.

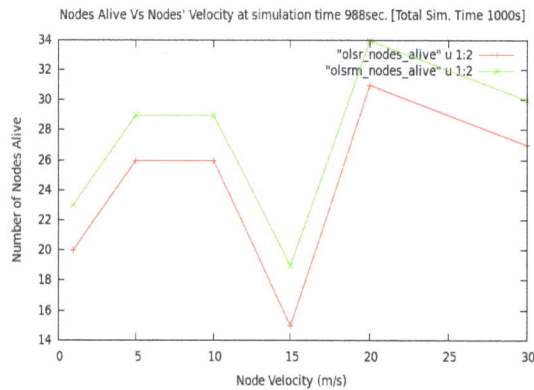

Nodes Alive Vs Nodes' Velocity at simulation time 988sec. [Total Sim. Time 1000s]

Fig. 2: Effect of node velocity on nodes alive.

B. Effect of Node Mobility and Node Density on Average End-to-End Delay:

For various Node's maximum velocity, OLSRM has less end-to-end delay, as multiple paths are available, than that of OLSR.

By varying node density, it has been observed that end-to-end delay is less for OLSRM than that of OLSR.

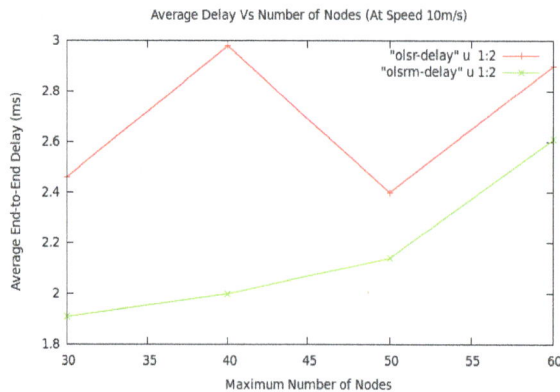

Average Delay Vs Number of Nodes (At Speed 10m/s)

Fig.3: Effect of node density on average end-to-end delay.

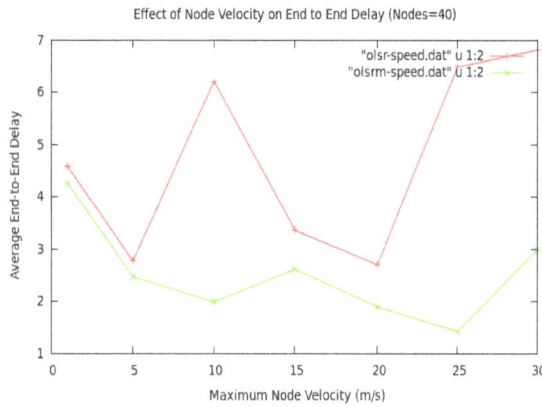

Fig.4: Effect of node velocity on average end-to-end delay

C. Effect of Node Mobility on normalized routing overhead:

To find the optimized energy path from the available source to destination multiple paths, it is expected that there will be increase in routing overheads compared to OLSR.

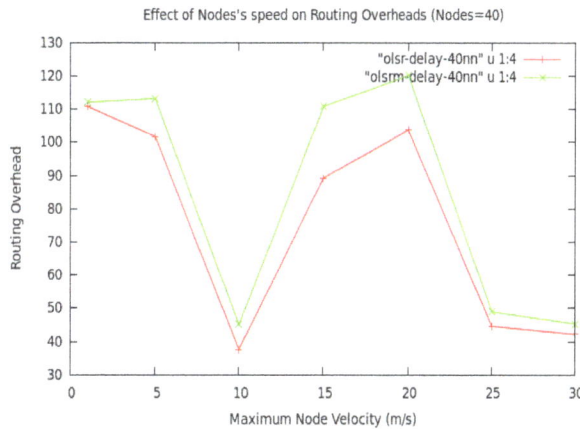

Fig.5: Effect of node velocity on normalized routing overheads

6. CONCLUSIONS

We examine the performance differences of OLSR and OLSRM. We measure Number of alive nodes and average end to end delay as QoS parameters.

OLSR, always uses shortest hop route, so congestion occurs and distribution of load is not considered. Also, OLSR does not consider available node energy of nodes for path selection and communication purposes. In this paper, algorithm for multipath OLSR with the addition of energy aware metric is given and simulation is performed using NS-2. Our simulation results show that OLSRM (modified OLSR with multipath) outperforms OLSR for number of alive nodes by 10 to 25% with considering performance parameters as node velocity and node density.

As expected, there is rise in routing overheads about 5-10% for node velocities up to 30 m/s.

Thus, congestion of the network disappears and load is transmitted uniformly throughout the network. The modified OLSR also gives the reduction in average end to end delay.

As a future work, we will evaluate optimum paths based on number of hops and available energy. Load will be mainly assigned to the main path, but if the energy of the intermediate nodes is reaching to threshold (given by the user and generally depends on data type), then another path to be considered. This will give the benefit of shortest hop route as well as optimum node energy consideration for longer life span of the network. Some methods or techniques to be added to reduce the normalized overheads.

ACKNOWLEDGEMENTS

We are thankful to Mr. Aniruddha B. Adoni as his help was very useful which gave a direction to complete the work by his timely critics and relevant inputs and suggestions.

REFERENCES

1. RASHMI: MANET WWW.SACHING.COM

2. Entan and Tania Jimenez, "NS Simulator for beginners", Lecture Notes, University of Merida, ESSI, Sophia-Antipolis, France. (December 4, 2003)

3. Chiang C., H. K. Wu, W. Liu, and M. Gerla, "Routing in clustered multihop mobile wireless networks with fading channel", Proc. IEEE SICON'97, PP.197-221(1997)

4. May Zin Do and Mazliza Othman "Performance comparisons of AOMDV and OLSR routing protocols for MANET", IEEE 2010. Iccea, vol. 1, pp.129-133(2010)

5. T. Clausen and P. Jacquet, "rfc3626.txt", Network working group DRAFT for OLSR protocol, Oct'03

6. Floriano De Rango and Macro Fotino "Energy Efficient OLSR performance Evaluation under Energy Aware Metrics", SPECTS 2009. IEEE 2009 pages: 193-198(2009)

7. Dongkyun Kim, J.J. Gracia-Luna-Aceves and Katia Obraczka, Juan-Carlos Cano and Pierto Manzoni,"Power-Aware routing based on the energy Drain Rate for Mobile AdHoc Networks", University of California, U.S.A., Polytechnic University of Valencia. IEEE 14-16 Oct., 2002; pages: 565-569(2002)

8. Floriano De Rango, Macro Fotino, Salvatore Marano, "EE-OLSR: Energy Efficient OLSR routing protocol for Mobile Adhoc Networks", D.E.I.S. Department, University of Calabria, Italy. IEEE, MILCOM 2008. 16-19 Nov., 2008. Pages: 1-7(2008)

9. Dimitrios J. Vergados, Athens and Nikkolas A. Patazis, "Enhanced route selection for energy efficiency in wireless sensor networks", Greece. ACM International Conference Proceeding Series; Vol 329, Article 63,2007

10. Bor-rong Chen and C. Hwachang, "Mobility Impact on Energy Conservation of Ad-hoc Routing protocols",Tufts University, Medford, USA. NSF Grant #0227879

11. Saoucene Mahfoudh, Pascale Minet, "An Energy Efficient routing based on OLSR in wireless ad hoc and sensor networks", INRIA, France. 22nd AINA workshop 2008, March 25-March 28. 2008 pp. 1253-1259(2008)

Permissions

All chapters in this book were first published in IJWMN, by AIRCC Publishing Corporation; hereby published with permission under the Creative Commons Attribution License or equivalent. Every chapter published in this book has been scrutinized by our experts. Their significance has been extensively debated. The topics covered herein carry significant findings which will fuel the growth of the discipline. They may even be implemented as practical applications or may be referred to as a beginning point for another development.

The contributors of this book come from diverse backgrounds, making this book a truly international effort. This book will bring forth new frontiers with its revolutionizing research information and detailed analysis of the nascent developments around the world.

We would like to thank all the contributing authors for lending their expertise to make the book truly unique. They have played a crucial role in the development of this book. Without their invaluable contributions this book wouldn't have been possible. They have made vital efforts to compile up to date information on the varied aspects of this subject to make this book a valuable addition to the collection of many professionals and students.

This book was conceptualized with the vision of imparting up-to-date information and advanced data in this field. To ensure the same, a matchless editorial board was set up. Every individual on the board went through rigorous rounds of assessment to prove their worth. After which they invested a large part of their time researching and compiling the most relevant data for our readers.

The editorial board has been involved in producing this book since its inception. They have spent rigorous hours researching and exploring the diverse topics which have resulted in the successful publishing of this book. They have passed on their knowledge of decades through this book. To expedite this challenging task, the publisher supported the team at every step. A small team of assistant editors was also appointed to further simplify the editing procedure and attain best results for the readers.

Apart from the editorial board, the designing team has also invested a significant amount of their time in understanding the subject and creating the most relevant covers. They scrutinized every image to scout for the most suitable representation of the subject and create an appropriate cover for the book.

The publishing team has been an ardent support to the editorial, designing and production team. Their endless efforts to recruit the best for this project, has resulted in the accomplishment of this book. They are a veteran in the field of academics and their pool of knowledge is as vast as their experience in printing. Their expertise and guidance has proved useful at every step. Their uncompromising quality standards have made this book an exceptional effort. Their encouragement from time to time has been an inspiration for everyone.

The publisher and the editorial board hope that this book will prove to be a valuable piece of knowledge for researchers, students, practitioners and scholars across the globe.

List of Contributors

Labib Francis Gergis
Misr Academy for Engineering and Technology Mansoura, Egypt

Nilesh P. Bobade
Department of Electronics Engineering, Bapurao Deshmukh COE, Sevagram, Wardha, M.S., India

Nitiket N. Mhala
Department of Electronics Engineering, Bapurao Deshmukh COE, Sevagram, Wardha, M.S., India

Maher HENI
Innovation of COMmunicant and COoperative Mobiles Laboratory, INNOV'COM Sup'COM, Higher School of Communication, Ariana, Tunisia

Ridha BOUALLEGUE
Innovation of COMmunicant and COoperative Mobiles Laboratory, INNOV'COM Sup'COM, Higher School of Communication, Ariana, Tunisia

Sunita Prasad
Centre for Development of Advanced Computing, NOIDA, India

Zaheeruddin
Department of Electrical Engineering, Jamia Millia Islamia, India

Pravanjan Das
Ericsson India Global Services Pvt. Ltd., Salt Lake, Kolkata, India

Sumant Kumar Mohapatra
Trident Academy of Technology, Bhubaneswar, Odisha

Biswa Ranjan Swain
Trident Academy of Technology, Bhubaneswar, Odisha

B.N Umesh
Research Scholar

Dr G Vasanth
Professor & Head, Dept of Computer Science, Govt Engg College, K.R Pate

Dr Siddaraju
Professor & Head, Dept of Computer Science, Dr. AIT, Bangalore

Natarajan Meghanathan
Jackson State University, 1400 Lynch St, Jackson, MS, USA

Anika Aziz
Applied Physics, Electronics & Communication Eng, University of Dhaka, Dhaka, Bangladesh

Shigeki Yamada
National Institute of Informatics, 2-1-2 Hitotsubashi, Chiyoda-ku, Tokyo 101-8430, Japan

Mahadev A. Gawas
Department of Computer Science BITS PILANI K.K. Birla Goa campus

Lucy J.Gudino
Department of Computer Science BITS PILANI K.K. Birla Goa campus

K.R. Anupama
Department of EEE/EI BITS PILANI K.K. Birla Goa campus

Joseph Rodrigues
Department of Electronics ATEC Verna goa

Gaurav Khandelwal
Birla Institute of Technology & Science-Pilani Hyderabad Campus, Hyderabad, INDIA

Giridhar Prasanna
Birla Institute of Technology & Science-Pilani Hyderabad Campus, Hyderabad, INDIA

Chittaranjan Hota
Birla Institute of Technology & Science-Pilani Hyderabad Campus, Hyderabad, INDIA

Dr.B.Vinayaga Sundaram
Department of Information Technology, Anna University, Chennai

G Rajesh
Department of Information Technology, Anna University, Chennai

A Khaja Muhaiyadeen
Department of Information Technology, Anna University, Chennai

R Hari Narayanan
Department of Information Technology, Anna University, Chennai

C Shelton Paul Infant
Department of Information Technology, Anna University, Chennai

G Sahiti
Department of Information Technology, Anna University, Chennai

R Malathi
Department of Information Technology, Anna University, Chennai

S Mary Priyanga
Department of Information Technology, Anna University, Chennai

Amjad Ali
School of Electrical Engineering and Computer Sciences National University of Science and Technology, Pakistan

Muddesar Iqbal
Faculty of Computer Science & Information Technology, University of Gujrat, Pakistan

Adeel Baig
School of Electrical Engineering and Computer Sciences National University of Science and Technology, Pakistan

Xingheng Wang
College of Engineering, Swansea University, Swansea, UK

Venetis Kanakaris
Department of Electronic and Computer Engineering, University of Portsmouth, Anglesea Road, Portsmouth, PO1 3DJ, United Kingdom

David Ndzi
Department of Electronic and Computer Engineering, University of Portsmouth, Anglesea Road, Portsmouth, PO1 3DJ, United Kingdom

Kyriakos Ovaliadis
Department of Electronic and Computer Engineering, University of Portsmouth, Anglesea Road, Portsmouth, PO1 3DJ, United Kingdom

Humaira Nishat
Department of Electronics and Communication Engineering, CVR College of Engineering, Hyderabad, India

Sake Pothalaiah
Department of Electronics and Communication Engineering, MRP UGC, Jawaharlal Nehru Technological University, Hyderabad, India

Dr. D.Srinivasa Rao
Department of Electronics and Communication Engineering, Jawaharlal Nehru Technological University, Hyderabad, India

Hassanali Nasehi
Department of Engineering, East Tehran Branch, Islamic Azad University (IAU) Tehran, Iran

Nastooh Taheri Javan
Department of Engineering, East Tehran Branch, Islamic Azad University (IAU) Tehran, Iran

Amir Bagheri Aghababa
Department of Engineering, East Tehran Branch, Islamic Azad University (IAU) Tehran, Iran

Yasna Ghanbari Birgani
Department of Industrial Engineering, Tarbiat Modares University (TMU) Tehran, Iran

P. Manickam
Department of Applied Sciences, Sethu Institute of Technology, India

T. Guru Baskar
Department of Applied Sciences, Sethu Institute of Technology, India

M.Girija
Department of Computer Science, The American College, India

Dr.D.Manimegalai
Department of Information Technology, National Engineering College, India

Kirti Aniruddha Adoni
Modern College of Engineering, Shivaji Nagar, University of Pune, , India

Radhika D. Joshi
College of Engineering, University of Pune, India